Extinction Rates

Extinction Rates

Edited by

JOHN H. LAWTON

NERC Centre for Population Biology
Imperial College, Silwood Park, UK

and

ROBERT M. MAY

University of Oxford and
Imperial College of Science, Technology, and Medicine, UK

Oxford · New York · Tokyo
OXFORD UNIVERSITY PRESS
1995

Oxford University Press, Walton Street, Oxford OX2 6DP

Oxford   New York
Athens   Auckland   Bangkok   Bombay
Calcutta   Cape Town   Dar es Salaam   Delhi
Florence   Hong Kong   Istanbul   Karachi
Kuala Lumpur   Madras   Madrid   Melbourne
Mexico City   Nairobi   Paris   Singapore
Taipei   Tokyo   Toronto
and associated companies in
Berlin   Ibadan

Oxford is a trade mark of Oxford University Press

Published in the United States
by Oxford University Press Inc., New York

A catalogue record for this book is available from the British Library

Library of Congress Cataloging in Publication Data
(Data applied for)
ISBN 0 19 854829 X

Typeset by Cotswold Typesetting Ltd, Gloucester

Printed in Great Britain on acid-free paper by
St. Edmundsbury Press
Bury St. Edmunds, Suffolk

Preface

Hardly a day passes without one being told that tropical deforestation is extinguishing roughly one species every hour, or maybe even one every minute. Such guesstimates rest on approximate species–area relations, along with assessments of current rates of deforestation and guesses at the global total number of species (which range from 5 to 80 million or more). While such figures arguably have a purpose in capturing public attention, there is a clear and increasing need for better estimates of impending rates of extinction, based on a keener understanding of extinction rates in the recent and far past, and on the underlying ecological and evolutionary causes.

To this end, we have brought together a coherently organized set of chapters, whose focus is firmly empirical. Indeed, with the exception of the introductory chapter, the book contains no mention of species–area relations. We believe the book provides a more wide-ranging and data-driven treatment of current and likely future extinction rates than has previously been drawn together in one place. In particular, we have aimed to identify unresolved issues and important questions.

In the opening chapter, May, Lawton, and Stork provide an overview by considering some of the main messages and questions that are beginning to emerge from increasingly detailed studies of past extinction events, on timescales ranging from hundreds of millions of years to the past few hundred years and less. This chapter surveys some of the differences among estimated extinction rates for different taxonomic groups (many of which are treated in more detail in later chapters), and discusses some of the ecological and evolutionary factors which may shape these differences. The opening chapter also gives a brief account of three different approaches to estimating future extinction rates (one based on species–area relations, and the other two based in different ways on the International Union for the Conservation of Nature's current categories of species 'endangerment'); these theoretical estimates of impending extinction rates are discussed in relation to the strictly empirical information about current and recent rates presented in subsequent chapters.

The next group of four chapters (2–5) sets the scene by reviewing the historical record of species extinctions, at different temporal scales, from millions to a few hundred years. Jablonski gives an overview of the sweep of the fossil record over the past 600 million years. Jackson looks at the major extinctions of species of molluscs, corals, and planktonic foraminiferans which occurred throughout tropical America about two million years ago in apparent response to the onset of glaciation in the Northern Hemisphere, and he compares this to the much less severe effects associated with subsequent glacial cycles, and associated temperature fluctuations and sea-level changes. Coope considers the history of fossil insects—mostly beetles—over the past few million years, and finds that the great climate oscillations associated with glacial cycles do not appear to have resulted in high rates of extinction among Quaternary fossil insects. Pimm, Moulton, and Justice examine the extinctions of bird species on Pacific Islands, beginning with the first waves of human colonists from 4000 to 1000 years ago, and accelerating more recently with the arrival of Europeans; these authors draw interesting comparisons among the Hawaiian Islands, the Marianas, New Caledonia, and the islands of South East Polynesia. A series of chapters (6–9) by Greuter, Bibby, Thomas and Morris, and Bond then deals with present and recent past patterns of extinction and endangerment among Mediterranean plants, the world's birds, British invertebrates, and the Cape flora of South Africa, respectively. In so doing, these authors expose significant differences in the patterns for different groups and different places, and they also raise fundamental issues about the causes and significance of endemism, mutualistic associations (as between plants and their pollinators or seed dispersers), and other ecological factors.

The next four chapters (10–13) build on the earlier ones, further to develop specific aspects of the ecological and evolutionary principles that underpin both understanding of extinction rates and the design of effective measures of conservation. Lawton examines the dynamics of populations, seeking to elucidate some basic constraints and rules that may make some species more prone to extinction than others. Most at risk are species with small geographical ranges and small total population sizes; since these two properties are commonly correlated, many such species are in double jeopardy. And—although there are many exceptions—tropical species are more likely to possess these adverse characteristics than are temperate or boreal ones. Other broad correlates of extinction include large body size and positions high in trophic webs, but here generalizations are harder to make and more difficult to understand. Nee, Holmes, May, and Harvey discuss how the accelerating avalanche of data on molecular phylogenies can be used to draw sharper inferences about past extinction patterns and processes. Specifically, they show how molecular phylogenies can provide information about the extent to which particular clades are likely to be under threat from extinction, how superficial or intuitive analyses of such phylogenies can often lead to wrong conclusions about the underlying evolutionary processes, and how different evolutionary processes can leave characteristic signatures upon the structure of reconstructed phylogenies. Margules and Austin deal with biological models for monitoring species' decline and with the construction of databases; they go on to show how these ideas have been put to practical use in optimising the design of conservation areas and refuges. Mace explains the steps that are being taken toward an increasingly objective classification of endangered species, according to estimates of their degree of extinction threat. She also indicates how such probability estimates for particular species—from ten families of reptiles, birds, and mammals—can be used to make very rough projections of impending extinction rates for members of these ten families (this new idea is one of the three methods of estimating future extinction rates, referred to above).

In the final chapter (14), Ehrlich goes to the heart of current problems, with an analysis of the continuing growth in human populations, and associated resource consumption (itself very different from place to place), which ultimately is the engine driving the extinction machinery, in a way that has no precedent in the history of life on Earth. Ehrlich sketches a possible path to a sustainable future, not just for us but for the biological diversity we are heirs to.

The present book derives from a Royal Society Discussion Meeting on 'Estimating extinction rates', held in October 1993 and organized by the two of us. We are indebted to the Royal Society and to the Linnean Society for making this stimulating meeting possible. The chapters in this book are, however, in most cases markedly different from' the papers published in the *Philosophical Transactions of the Royal Society* (1994) **344**, 1–104 after the meeting: in several places, the book chapters offer a broad view of their topic, whereas the *Philosophical Transactions* version complements the chapter by confining itself mainly to technical details of new methods or supporting material; the opening chapter here, and significant parts of other chapters, where written specifically for this book. We are much indebted to all the authors, who cheerfully and punctually re-wrote their papers for this book on a tight schedule. We are also grateful to the editorial staff at Oxford University Press for all their help.

John H. Lawton and Robert M. May
February 1994

Contents

Contributors

M. P. Austin   CSIRO Division of Wildlife and Ecology, PO Box 84, Lyneham, ACT 2602, Australia

Colin J. Bibby   BirdLife International, Wellbrook Court, Girton Road, Cambridge CB3 0NA, UK

W. J. Bond   Botany Department, University of Cape Town, Private Bag, Rondebosch 7700, South Africa

G. R. Coope   Department of Geography, Royal Holloway, University of London, Egham, Surrey TW20 0EX, UK

Paul R. Ehrlich   Center for Conservation Biology, Department of Biological Sciences, Stanford University, Stanford CA 94305, USA

Werner Greuter   Botanischer Garten und Botanisches Museum Berlin-Dahlem, Königin-Luise-Strasse, 6–8, D-14191 Berlin, Germany

Paul H. Harvey   Department of Zoology, University of Oxford, South Parks Road, Oxford OX1 3PS, UK

Edward C. Holmes   Department of Zoology, University of Oxford, South Parks Road, Oxford OX1 3PS, UK

David Jablonski   Department of Geophysical Sciences, University of Chicago, 5734 South Ellis Avenue, Chicago IL 60637, USA

J. B. C. Jackson   Smithsonian Tropical Research Institute, Box 2072, Balboa Republic of Panama. *Postal address*: Smithsonian Tropical Research Institute, Unit 0948, APO AA 34002-0948, USA

Lenora J. Justice   Department of Biology, Georgia Southern University, Statesboro, GA 30460, USA

John H. Lawton   NERC Centre for Population Biology, Imperial College, Silwood Park, Ascot, Berks SL5 7PH, UK

Georgina M. Mace   Institute of Zoology, Zoological Society of London, Regents Park, London NW1 4RY, UK

C. R. Margules   CSIRO Division of Wildlife and Ecology, PO Box 84, Lyneham, ACT 2602, Australia

Robert M. May   Department of Zoology, University of Oxford, South Parks Road, Oxford OX1 3PS, UK

M. G. Morris   Institute of Terrestrial Ecology, Furzebrook Research Station, Wareham, Dorset BH20 5AS, UK

Michael P. Moulton   Department of Biology, Georgia Southern University, Statesboro, GA 30460, USA. *Current address*: Department of Wildlife and Range Sciences, 118 Newins-Ziegler Hall, University of Florida, Gainsville, FL 32611, USA

Sean Nee   Department of Zoology, University of Oxford, South Parks Road, Oxford OX1 3PS, UK

Stuart L. Pimm   Department of Zoology and Graduate Program in Ecology, The University of Tennessee, Knoxville TN 37996, USA
Nigel E. Stork   Department of Entomology, The Natural History Museum, Cromwell Road, London SW7 5BD, UK
J. A. Thomas   Institute of Terrestrial Ecology, Furzebrook Research Station, Wareham, Dorset BH20 5AS, UK

1

Assessing extinction rates

Robert M. May, John H. Lawton, and Nigel E. Stork

1.1 Introduction

In this opening chapter, we review what we see as some of the major themes, and major questions, which emerge from the data presented in the subsequent chapters, all of which are empirically-oriented.

We also sketch the essential elements of three different theoretical approaches to estimating likely future rates of extinction. The first, and by far the most familiar, uses species–area relations in combination with current or projected rates of tropical deforestation. The second method provides a very rough estimate, based on the current rates at which species in better-studied groups (specifically, birds, mammals, and palm trees) are 'climbing the ladder' of the International Union for the Conservation of Nature and Natural Resources' (IUCN) categories of threat, from 'vulnerable' to 'endangered' to 'probably extinct' to certified extinction. The third method is a new one, presented here in Chapter 13 by Georgina Mace; it uses the estimated probabilities of extinction, as functions of time, which underpin the IUCN categories of threat for individual species. In reviewing these three approaches, which give surprisingly concordant answers, we emphasize the many uncertainties and other problems associated with them. We also discuss the discrepancies between rates of documented extinction over the past century or so, and the roughly hundredfold higher rates of projected future extinctions.

Our chapter concludes with some comments on the implications for conservation of biological diversity which emerge from the later chapters, and from other recent work on patterns of commonness and rarity among different groups and different places.

1.2 Extinction rates in the fossil record

Biological diversity has been increasing from at least 3.5 billion years ago (3.5×10^9 ya), the date assigned to the oldest objects that are probably microbial fossils (Awramik *et al.* 1983). There is as yet no agreed explanation of why multicellular organisms apparently did not begin to diversify until around 1.4 billion years ago (Walter *et al.* 1990), nor why multicellular animals initiated their explosive diversification in the Early Cambrian, some 600 million years ago (600 mya). Plants and animals did not invade the land and begin their pattern of terrestrial diversification until somewhat later still, around 450 mya. That is, these latter three milestones—the beginning of diversification of multicellular organisms, multicellular animals, and terrestrial plants and animals—fall around 60%, more than 80%, and almost 90% along the road from the beginning of life on earth to the present, respectively. As Sepkoski (1992) emphasizes: 'The basic observation here is that diversification has not been continuous. This is not a conclusion that could easily be reached from first evolutionary principles'.

In dealing with past extinction rates, our attention—both in this opening chapter and throughout the book—is confined almost exclusively to the plant and animal kingdoms, as revealed by the fossil and other records since the Cambrian. We recognize that communities of soft-bodied multicellular organisms undoubtedly pre-date the Cambrian, but it is difficult to estimate their diversity. We also recognize that arguably the most important and interesting phase in the evolution of life on earth was the first two billion years or so, from the emergence of simple self-replicating molecules to the first prokaryotes. Last and perhaps most important, we recognize that microorganisms, representing five to seven recognized kingdoms of organisms, and including viruses, bacteria, protozoans, and algae, created the planet's oxygen-rich atmosphere, and that the 'ecosystem services' they provide today as decomposers, soil maintainers, and so on, underpin all other living things. Be all this as it may, our focus is on only two kingdoms, plants and animals, over the past 600 million years.

Table 1.1 summarizes a variety of estimates that have been made of the average lifespan of a species in the fossil record, from origination to extinction. Many of these estimates were collected together by Raup (1978), whose own assessment of around 11 million years for the average lifespan of invertebrate species is based on computer analysis of some 8500 cohorts of fossil genera. The other estimates summarized in Table 1.1 are of varying degrees of sophistication; some of the subtleties and difficulties in such analyses are discussed in more detail in Chapters 2 and 11. All in all, Table 1.1 suggests the average species has a lifespan of around 5–10 million years.

It follows that, if the average species lives 5–10 million years, and the duration of the fossil record we are considering is around 600 million years, then—all things being equal—the Earth's current tally of plant and animal species represents about 1–2% of all those that have ever lived.

Of course, all things are not equal. Figure 1.1 shows the numbers of families of marine animals as a function of time, throughout the Phanerozoic (which is essentially the 600 million year span we are focused on). As Sepkoski (1992) emphasizes, the patterns in diversity at the level of families seen in Fig. 1.1 are broadly in agreement with the corresponding patterns at higher (orders) and lower (genera, species) taxonomic levels; in particular, they correlate well with underlying species diversity (Sepkoski et al. 1981; Bambach and Sepkoski 1992). Moreover, similar 'stair-step' patterns have been documented—although with different time axes—for terrestrial plants (Niklas et al. 1983; Knoll 1986), tetrapod vertebrates (Benton 1985, 1990), and possibly for terrestrial arthropods (Sepkoski and Hulver 1985; Labandeira and Sepkoski 1993; see Fig. 1.2 and further discussion, below). Figure 1.1 thus suggests that the diversity of life-forms

Table 1.1 Estimates of species' lifespan, from origination to extinction

Taxon	Source of estimate	Species' average lifespan (mya)
All invertebrates	Raup (1978)	11
Marine invertebrates	Valentine (1970)	5–10
Marine animals	Raup (1991)	4
Marine animals	Sepkoski (1992)	5
All fossil groups	Simpson (1952)	0.5–5
Mammals	Martin (1993)	1
Cenozoic mammals	Raup and Stanley (1978)	1–2
Diatoms	Van Valen (1973)	8
Dinoflagellates	Van Valen (1973)	13
Planktonic foraminifera	Van Valen (1973)	7
Cenozoic bivalves	Raup and Stanley (1978)	10
Echinoderms	Durham (1970)	6
Silurian graptolites	Rickards (1977)	2

in the sea—measured at taxonomic levels ranging from orders to species—rose abruptly throughout the Cambrian, to attain a rough plateau toward the end of the Ordovician; this plateau continued, albeit with significant fluctuations, until the great wave of Permian extinctions (which marked the end of the Palaeozoic), since when there have been fluctuations about a roughly steady upwards trend. Sepkoski (1992) concludes that present marine diversity is roughly twice the average level over the entire span of the Phanerozoic (and that the biases introduced by better knowledge of more recent events and records is not a significant factor in this conclusion).

Today, however, only about 15% of recorded species are found in the sea; the majority are terrestrial (Barnes 1989; May 1994). If this ratio of recorded species accurately reflects reality (and some would argue it does not: Grassle and Maciolek 1992, but see Hammond 1992 and May 1992, 1994), and if it has been roughly thus since the land was colonized, then we must focus primarily on terrestrial patterns. As noted by Sepkoski (1992) and others (see the preceding paragraph), terrestrial diversity of plants and animals, again measured at taxonomic levels ranging from species to orders, shows patterns very roughly similar to those in Fig. 1.1, but on a shorter timescale of around 450 million years. Terrestrial diversity rose steadily throughout the Devonian, to level out in the Carboniferous and Permian; the trends since the Triassic, accentuated in the Tertiary, have been toward increasing diversity (owing, in part, to the Earth's land masses becoming more fragmented and dispersed toward the end of the Tertiary than at any other point in the Phanerozoic; the degree to which this helps explain corresponding increases in marine diversity is less clear). In short, Sepkoski's (1992) rough rule, that the current marine diversity is about twice the Phanerozoic average, also suggests that the larger terrestrial diversity today is about twice the average over the 450 million year history of life on land.

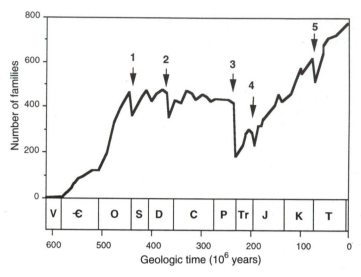

Fig. 1.1 The history of the diversity of marine animal families throughout the Phanerozoic, as a function of time. The curve connects 77 discrete data points, each showing the total number of well-skeletonized families known from a particular stratigraphic stage. The arrows labelled 1 to 5 identify the five major events of mass extinction: 1, end Ordovician (O); 2, late Devonian (D); 3, end Permian (P); 4, end Triassic (Tr); 5, end Cretaceous (K). The durations of the various geological periods are indicated on the time axis. In addition to the five just given, the abbreviations for the remaining periods are: V, Vendian; ᴲ, Cambrian; O, Ordovician; S, Silurian; D, Devonian; C, Carboniferous; P, Permian; Tr, Triassic; J, Jurassic; K, Cretaceous; T, Tertiary. (After Sepkoski 1992.)

Putting all this together (and recognizing that the difference between 600 and 450 million years is relatively insignificant, given the accuracy of our estimates), we double our 'all things being equal' estimate, to conclude that today's species of plants and animals constitute roughly 2–4% of the total there have been.

These figures, of course, are sensitive to our estimates of the average species' lifespan inferred from the fossil record. But the actual numbers of fossil species that have been named and recorded (as distinct from the numbers inferred from our more-complete knowledge of higher-order taxa, such as families and orders) is a small sample of the actual total. Basing his analysis on an earlier one by Raup (1976), Sepkoski (1992) estimates that some 250 000 species of fossil marine animals have been documented. This is roughly equal to his estimate of 200 000 for the number of animal species in the ocean today. Hence, by essentially the same arguments as we have just used to assess the fraction of all species that are alive today, Sepkoski concludes that the known fossil record of marine species represents a sample of only 2–4% of the total (in fact, he obtains 2%, because he takes the average species' lifespan to be 5 million years, at the lower end of our 5–10 million year range).

Actually, the overall sample is worse than this. Around 95% of all fossil species are marine animals (Raup 1976), partly because a lot of them are shallow-sea creatures that dwelt in the kinds of environments which produce rich beds of fossils. But we just noted that around 85% of today's recorded plant and animal species are terrestrial. If a similar proportion of species were terrestrial over a reasonable fraction of the Phanerozoic, then documented fossil species are unlikely to represent as much as a sample of 1% of the true total.

They are, moreover, a heavily biased sample, being predominantly species of marine invertebrates. Estimated average species' lifespans, however, vary significantly among groups, as Table 1.1 shows (and as is discussed more fully in the next four chapters). In particular, the average lifespan of mammal species in the fossil record is around 1 (or possibly 2) million years, which is an order-of-magnitude less than the roughly 10 million year estimate by Raup (1978) for invertebrates (essentially all of which are marine). Even among marine invertebrates, there are substantial differences from group to group, with Mesozoic ammonoids, for example, having average species' lifespans of only 1–2 million years (Kennedy 1977). Conversely, there are indications (in Chapter 4 by Coope, and in Fig. 1.2 below) that insect species may be unusually long-lived, at least in north-temperate regions. In addition, Labandeira and Sepkoski (1993) have suggested that the record of insect diversification is one of roughly linear increase from low levels around 300–400 mya; today, insects account for 56% of all named and recorded species (indeed, to a first approximation, all contemporary species are insects). Such a combination, for insects, of relatively long species' lifespan and numerical preponderance within recent species totals could seriously undercut all the above estimates. Even if we make the modest assessment that the total number of insect species on Earth today (recorded plus yet-unknown) is only around 3 million or so, it is conceivable that 10% or more of all plant and animal species ever to have lived are with us today, and they are terrestrial insects.

Finally, notice that our 'all things being equal' estimate implies that, on average, the plant and animal species living at any one time during the Phanerozoic represent 1–2% of the overall total. The five great 'spasms' of extinction in the fossil records are indicated in Fig. 1.1. Those which marked the ends of the Ordovician, Devonian, Triassic, and Cretaceous (marked 1, 2, 4, 5, respectively, in Fig. 1.1) eliminated 65–85% of animal species in the ocean, while that at the end of the Permian (marked 3 in Fig. 1.1) extinguished 95% or more of marine species. These five mass extinctions differed from each other, and each one affected different taxonomic groups differently (Benton 1986); for example, insect diversity was particularly affected by the extinction event at the end of the Permian (marked 3). Nevertheless, if we assume that terrestrial extinction levels very roughly parallelled marine ones, we conclude that the total fraction of all plant and animal species eliminated by the five mass extinction events (the 'Big Five') is 1–2% multiplied by $(4 \times 0.75 + 0.95)$, or roughly 4–8%. This accords with Raup's (1986, 1992) more detailed estimates that 90–96% of all extinctions occurred outside these major episodes.

This section has dealt exclusively with *proportions* of species to have become

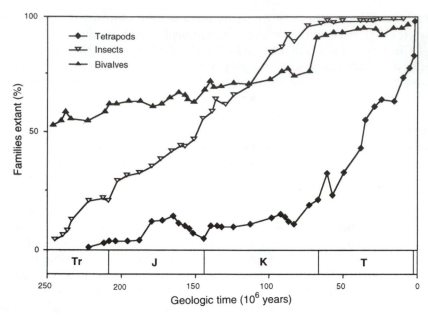

Fig. 1.2 These curves show currently-extant families as a proportion of the total number of families found in the fossil record, at various times over the past 250 million years. The curves are for insects, marine bivalves, and terrestrial vertebrate tetrapods, as indicated. The data are plotted for discrete stratigraphic stages, and the abbreviations for the geological periods are as in Fig. 1.1. The main features of these three different curves are discussed in the text (notice that the curves need not be monotonic, because radiations of short-lived families can temporarily depress the fraction that survive into the present). In particular, these data suggest that essentially no insect families have become extinct over the past 100 my or so. (After Labandeira and Sepkoski 1993.)

extinct, and so on. Any assessment of *absolute* rates of species extinction requires an estimate of the total number of plant, animal, and other species currently extant, and thence (by the kinds of arguments used above) ever to have lived. A full discussion of guesstimates for such species totals, and in particular any discussion of recent (and sometimes dramatic) suggestions for upward revisions for particular groups, would require its own chapter (for a short summary, see May 1990*a*, 1994; Hammond 1992; or Stork 1988). We content ourselves with the estimate of 5–10 million species currently alive. Combined with the estimate of 5–10 million years for the average species lifespan, this gives a very rough estimate for the 'background' extinction rate in the fossil record of one species per year (but it could easily be at least a factor two bigger or smaller).

1.3 Ecological and evolutionary factors affecting past extinctions

The following chapters discuss some of the patterns discernible in which species survive, and which do not, at various times in the fossil record.

Jablonski and Jackson both survey evidence in support of the view that species which are fairly widely dispersed, and particularly those that achieve this by wide dispersal of larval stages, are more likely to survive extinction events (both the 'Big Five' and the many smaller events, on nested hierarchies of scales), than are more sessile or range-restricted species. This is, indeed, the prevailing view (e.g., Diamond 1984; Simberloff 1986a), although counter-examples can be found. Vermeij (1993), for instance, examined a collection of 15 marine species (1 gastropod, 1 coral, 3 mammals, 10 birds) that have disappeared during the human-dominated post-Pleistocene period: although 9 of these did have narrow ranges (being known from a single island or a single marine biogeographical province), at least 5—one-third of the total—have large ranges, embracing parts of two or more marine biogeographical provinces. On balance, we believe the conventional generalization is valid, although it must be applied with caution to specific cases.

This generalization has recently received theoretical support, which goes beyond the commonsense observation that you are obviously better off if you are widespread and abundant than if you are localized and rare. Tilman *et al.* (in press) have generalized earlier studies by Nee and May (1992), to examine multispecies metapopulations in which a number of species can persist together in an environment where there are many habitat patches: metapopulations of each species maintain themselves by a balance between local extinctions and colonizations; there is a hierarchy of competitive ability, so that competitively superior species exclude competitively inferior ones from co-occupied patches; and inferior competitors persist by virtue of higher colonization rates and/or lower patch-mortality rates. This system has complicated dynamics, and exhibits inherent 'limits to similarity' (manifested through the colonization and mortality parameters, and not arbitrarily imposed as in many older studies of this subject). If this metaphorical system is stressed by environmental events which reduce the total number of habitat patches, or which reduce the overall biomass in the system, it is the superior competitors (often the most abundant species) which tend to be lost, while the superior dispersers survive. This is, admittedly, an over-simplified metaphor, but it has interesting implications (Tilman *et al.* in press).

Another theme which emerges in several of the later chapters of this book is that environmental changes—human-related or otherwise—are likely to have less effect on floras and faunas which have undergone similar stresses in the not-so-far-distant past. Thus, Jackson shows the onset of glaciation in the northern hemisphere around 2 mya caused dramatic extinctions among species of molluscs, corals, and planktonic foraminiferans throughout tropical America, whereas subsequent cycles of glaciation, temperature fluctuations, and sea-level changes have had comparatively little effect. Pimm, Moulton, and Justice show that, for birds, there are fewer recent extinctions, and fewer currently-endangered species, on the islands of the western Pacific (which were colonized by humans some 3000–5000 ya) than on the more recently colonized Hawaiian and other eastern Pacific islands. Greuter compares records of recent plant extinctions for 'Mediterranean' floras in the actual Mediterranean, South America, California, and Western Australia. These extinction rates are lowest (around 0.1%) for the Mediterranean, where human impacts are oldest, and highest (around 1%, and possibly higher) in Western Australia, where they are most recent. All this points to the conclusion that species assemblies which have been winnowed by relatively recent vicissitudes (on timescales from thousands to millions of years) are likely to suffer less from subsequent impacts of the same general kind. Expressed this way, the conclusion is unsurprising. But taken in isolation, the relatively low rate of recent extinction, and perhaps even of current endangerment (around 15%), for the flora of the Mediterranean littoral—one of the world's most human-impacted regions—is surprising.

Various other suggestions have been made about factors which may predispose particular species or groups to extinction. For several of these factors, no clear message emerges from the data surveyed in the book. Thus, Jablonski finds that, over the fossil record, position in the trophic hierarchy does not correlate clearly with extinction; top predators do not appear to have generally higher extinction rates. Lawton shows that the effects of body size upon extinction rates are complex, with no simple rules.

On the other hand, some kinds of trophic links do have implications for extinction processes. In Chapter 9, Bond examines the ways in which plant species can be put at risk by loss of pollinators or seed dispersers. He considers the trade-offs among different categories of risk, and shows that plants often compensate for high risk in one aspect of their relation with a mutualist by low risk in another aspect. For example, self-incompatible plants with specialized pollinators that are rare often have vegetative propagation. Bond notes that some systems—including the South African Cape flora and lowland tropical rainforests—lack this kind of compensation, and are thereby especially vulnerable to 'knock-on' extinctions when mutualists are lost. This analysis raises subtleties that go well beyond the simple observation that if one partner in an obligatory dyad is extinguished, the other is doomed (even though it may take time to disappear if it is a long-lived tree: Janzen's 'living dead').

Some clear-cut co-extinction events do, however, go unnoticed, largely because little things in general, and parasites in particular, elicit scant concern. For instance, Stork and Lyal (1993) note that the mournful label attached to Martha, the last passenger pigeon, in the Smithsonian Institution's Natural History Museum makes no mention of at least two species of obligate ectoparasite, the lice *Columbicola extinctus* and *Campanulotes defectus*, which accompanied her into oblivion, nor does the World Conservation Monitoring Centre (WCMC 1992) list *any* species of Phthiraptera (lice) or Siphonaptera (fleas) as extinct. This kind of selective attention has larger implications (Gaston and May 1992).

1.4 Differences in extinction rates among groups

Coope's main theme, highlighted in his title for Chapter 4, is that species of northern insects in general, and British beetles in particular, have shown remarkably few extinctions over the past few million years. He suggests that this may be because the mobility of most insects allows them to track environmental changes, shifting their ranges as climate alters. One slight problem here is that, as Nix (1986), Busby (1986), and others have shown, climate changes do not always cause a species' range (as defined by physiological and other constraints) to move smoothly and continuously across the map; the result is often discontinuous fragmentations and saltatory jumps in a mosaic of potentially suitable habitat. Coope himself emphasizes that such environmental 'tracking' will not be possible if changes are too fast, too extreme, and/or on too large a scale, and he suggests this may reconcile the constancy he sees in the Pleistocene insect fauna of Britain with the list of recent losses (and the larger list of impending losses) surveyed by Thomas and Morris in Chapter 8.

The fact remains that, in this book, there is a notable difference between the patterns of relatively low extinction rates seen for northern insects by Coope and for 'Mediterranean' floras by Greuter, versus the relatively high rates documented for birds by Pimm *et al.* and by Bibby. Mammalian extinctions in the late Pleistocene, although not discussed elsewhere in the book, also tell a story different from Coope's insects: about 40% of the genera of large mammals in Africa, and more than 70% in North America, South America, and Australia, were extinguished over the past 100 000 years or less (the timescales are correlated with human arrivals in Australia and the New World: see Martin and Klein 1984).

These differences among different groups are writ larger in the estimates of the characteristic lifespans of species, from origination to extinction, summarized in Table 1.1 above. Figure 1.2 suggests the differences among groups may be even greater than indicated in Table 1.1 (Labandeira and Sepkoski 1993). This figure shows the families of insects, marine bivalves, and terrestrial vertebrate tetrapods that are alive today, as fractions of all families (in each group) found in the fossil record as we reach back over the past 250 million years or so. Admittedly most insect families are more speciose than vertebrate ones, and thereby possibly more resistant to extinction if species lifespans' are comparable (see Jablonski, Chapter 2), but even so the differences seen in Fig. 1.2 are remarkable. The differences between the average family lifespans of bivalves and tetrapods accords with the order-of-magnitude differences in species' lifespans between the two groups indicated in Table 1.1 (roughly 10 my versus 1 my). As mentioned earlier, Fig. 1.2 suggests that, over the past 100 my or so, the average lifespan of an insect species could be well in excess of 10 my.

Such differences among groups prompt questions about whether we are applying the same taxonomic criteria to all groups. There are problems with 'taxonomic inflation' over the years; what was a genus to Linnaeus might be a family, or even higher, today (Sepkoski 1992). These instabilities, applying differently to different groups and dependent to some degree on the amount of attention given to different groups, create problems when we try—as we have been doing above—to draw inferences about the history of particular groups of species from their superspecific taxonomy (Patterson and Smith 1989). Such methodological differences among taxonomic groups show up in other fundamental ways: for example, Selander (1985) has observed that different strains of what is currently classified as a single bacterial species, *Legionella pneumophila*, have nucleotide sequence homologies (as revealed by DNA hybridization) of less than 50%; this is as large as the characteristic genetic distance between mammals and fishes.

All this being said, the balance of current evidence suggests there are real differences in characteristic extinction rates among different groups, possibly grounded in systematic ecological differences (some of which may be caricatured as the differences between '*r*-selected' and '*K*-selected' species, to use a metaphor which has been much abused). But we must be wary about applying any such tentative generalizations to current situations. In particular, most of our knowledge of Paleozoic insects comes from what was then in the tropics (the Dominican amber samples insects of the subtropics, as do the Cretaceous ambers of Mexico), whereas most of our detailed knowledge of late Cenozoic insects is for non-tropical faunas (Coope, this volume).

Against this background of past events, we now turn to recent and likely future events.

1.5   Current and recent extinction rates

As discussed in more detail in Chapters 7 and 13, by Bibby and Mace, respectively, the information we have about documented extinctions of animal species since around 1600 has been drawn together by the IUCN (1990) and the WCMC (1992). The IUCN/WCMC Red Data Books also list species under various categories of extinction threat: 'possibly extinct', 'endangered' (survival unlikely if causal factors continue), 'vulnerable' (likely to become 'endangered' if current trends continue), along with 'rare' (but not necessarily threatened), 'status unknown', and 'not [or no longer] threatened'. A more analytic account of these categories, and of attempts to give greater precision to them over recent years, is in Chapter 13 (see also Mace 1994; and Mace and Lande 1991). The WCMC database also provides information about seed-bearing plants threatened with extinction, using comparable categories of threat.

Table 1.2 summarizes these data, showing both the absolute numbers and the proportions of major animal and plant taxa that have become extinct since 1600, or that are currently threatened. Roughly half these recorded extinctions of 485 animal and 585 plant species have taken place this century (Smith *et al.* 1993*b*).

Many authors, including Bibby and Mace in this book, have emphasized the many difficulties and shortcomings in the database summarized by Table 1.2. We outline some of these problems, because several raise interesting and wider questions.

First, notice that documented extinctions among insect species are two orders-of- magnitude less (as a fraction of their recorded fauna) than among vertebrates, over the past 400 years or so. This ratio is, of course, even smaller if one recognizes that the true number of insect species is almost surely at least 3 million,

Table 1.2 Species in major taxa that have become extinct since 1600 or are threatened with extinction. (After Smith *et al.* 1993*a*.)

	No. of species certified extinct since 1600	No. of species listed as threatened*	Approx. total of recorded extant species (in thousands)	Approx. extinct (%)	Approx. threatened (%)
Animals					
Molluscs	191	354	100	0.2	0.4
Crustaceans	4	126	40	0.01	0.3
Insects	61	873	1000	0.006	0.09
Vertebrates	229	2212	47	0.5	5
Fishes	29	452	24	0.1	2
Amphibians	2	59	3	0.1	2
Reptiles	23	167	6	0.4	3
Birds	116	1029	9.5	1	11
Mammals	59	505	4.5	1	11
Total (animals)	485	3565	1400	0.04	0.3
Plants					
Gymnosperms	2	242	0.8	0.3	30
Dicotyledons	120	17 474	190	0.06	9
Monocotyledons	462	4421	52	0.9	9
Palms	4	925	2.8	0.1	33
Total (plants)	584	22 137	240	0.2	9

*'Threatened' here includes IUCN categories of 'vulnerable', 'endangered', 'probably extinct', but does not include other categories of concern, such as 'rare' or 'insufficiently known'

and possibly much higher. Although we have argued, in the preceding section, that the average insect species' lifespan in the fossil record might be an order-of-magnitude longer than those of tetrapods, there is no suggestion of a disparity as great as seen in Table 1.2. Rather, the differences among taxa in the proportions listed as extinct or threatened result primarily from differences in the attention paid to them. For one thing, the ratio of taxonomists to numbers of species is roughly 100 times greater for vertebrates than for invertebrates, and 10 times greater for vertebrates than for vascular plants (Gaston and May 1992). For another thing, this differential attention is seen in the rates at which new species are recorded, and in the completeness of collections, for different groups: new bird species are found at the rate of around 3–5 per year (or 0.03–0.05% of the total in the group); tropical botanists can expect to find roughly one new plant species per 100 species collected (G.T. Prance, pers. comm.); but collections of insect species, or fungi, or marine macrofauna from previously unstudied locations have shown proportions of new species ranging from 20% to 50–80%, and occasionally even more (May 1990a, 1994; Hawksworth 1991; Poore and Wilson 1993). In short, some of the groups in Table 1.2 are much better-studied than others.

Second, even for the comparatively well-known groups, such as birds and mammals, the documented extinctions are surely underestimates (Diamond 1989; Pimm et al., Chapter 5). The stringent IUCN criteria required to be listed as 'extinct' in Table 1.2 are not easily satisfied by animals and plants in remote places. Diamond (1989), for example, studied the bird fauna of the Solomon Islands, where 164 species have been recorded; 12 of these have not been seen since 1953 or earlier, nor could he find them. These 12 bird species are almost certainly extinct, yet only 1 is so certified. Among vertebrates, freshwater fish species almost surely are proportionately the most endangered, as a result of pollution of rivers (although introduced species are also causing major problems, of which those in Lake Victoria may be the best known). One indicative recent four-year search for the 266 species of exclusively freshwater fishes recorded last century in lowland peninsula Malaysia found only 122 (Mohsin and Ambok 1983). Yet relatively few fish species show up as extinct or threatened in Table 1.2 (Wilson 1992).

These problems are exacerbated by so many species being known from only a single collecting site. Stork and Hine (in press), for example, estimate that such is the case for about 40% of the estimated 400 thousand recorded species of beetles. Under these circumstances, it is difficult to assess anything other than local extinctions.

Third, rates of documented extinction and threat vary among geographical locations (Smith et al. 1993b; Mawdsley and Stork 1994). Some of these patterns may reflect reality, but others surely do not. An example of a real pattern is that roughly 61% of the animal extinctions (but only 27% of the plants) in Table 1.2 come from island endemics, mainly in the Pacific Ocean (58% and 72% of all island extinctions for animals and plants, respectively). The differential vulnerability of previously-isolated island endemics to all four of Diamond's (1989) 'evil quartet'— habitat destruction and fragmentation, introduced species, over- exploitation, and chains of extinction or 'knock-on' effects—has been much discussed; the numerical preponderance of Pacific islands is probably because there are more of them. An example of an artefactual pattern is that about two-thirds of the recent animal extinctions recorded on continents are from North America and the Caribbean, and about one-fifth from Australia. The relative paucity of recorded animal extinctions from South America, Africa, and Asia surely reflects less knowledge and less recording effort from these species-rich regions (the recent discovery of three new species of primates in Brazil, and of a new bovid in Vietnam, highlights this: Mawdsley and Stork 1994). Similarly, roughly two-thirds of recorded continental plant extinctions are from North America and Australia. All 45 extinctions in sub-saharan Africa are from the Cape flora of South Africa; as Bond's Chapter 9 makes clear, this comparatively well-studied flora may have particular vulnerability, but it seems unlikely that no other African floras have suffered losses.

These various problems in the database summarized by Table 1.2 show up clearly—even laughably—when we look closely at its 61 extinct insect species (Mawdsley and Stork 1994). Of this 61 species, 33 are lepidopterans. The remaining 28 species are distributed among orders in rough proportion to the speciosity of orders, but the over-representation of lepidopterans in general, and butterflies in particular, most likely derives from their status as 'honorary birds'. Looking from another perspective, it is perhaps reasonable that 51 of the 61 species are island endemics. But 42 of this 51 are from Hawai'i. Of the 10 continental extinctions, 9 are from North America (specifically, the United States), 1 from Europe (specifically, Germany). Again, this speaks eloquently of where the researchers, rather than the endangered species, live.

It is illuminating to observe that for the insect and other invertebrate fauna of Britain, which is better-known than that of any other country, the rates of extinction and endangerment (within Britain) are comparable to those for birds and mammals (more generally) in Table 1.2. The British Red Data Books for insects (which covers some 14 000 of Britain's estimated 22 000 insect species) and for other non-marine invertebrates are summarized in Table 1.3. From these data, we see that roughly 1% of Britain's insects have been extinguished this century (although essentially all survive, and many are thriving, elsewhere), and 6% are currently threatened. Although there is only one recorded extinction among Britain's other invertebrates, a similar fraction, 7%, are listed as threatened. These proportions among a comparatively well- known invertebrate fauna are similar to the fraction of the world's birds and mammals on Red Lists of threatened species, strengthening the argument that many of the entries in Table 1.2 reflect lack of knowledge, not lack of endangerment.

Finally, we note that, for the comparatively well-studied birds and mammals, roughly 100 of the documented extinctions in Table 1.2 took place this century. There are around 14 000 species of birds and mammals in total, so this represents documented extinction of about 1% of these species over the past 100 years. Such a rate translates to an expected lifespan, before extinction, of around 10 000 years for the average bird or mammal species. Although this may seem long, it is 2 to 3 orders-of-magnitude shorter than the average species lifespan of 1–10 my deduced from the fossil record.

1.6 Estimated future extinction rates from species–area relations

Essentially, all projections of impending rates of extinction are based on species–area relations, combined with estimated rates of loss of habitat ('area') due to deforestation or other processes. There have been many such projections (for reviews, see the papers in Whitmore and Sayer 1992), but they all have the same simple basis.

The species–area relation is an empirical rule, based on a variety of studies of how the number of species, S, of a particular taxonomic group (beetles, birds, vascular plants, etc.) found on individual islands within an archipelago depends

Table 1.3 Insect and other non-marine invertebrate species that have become extinct, or are threatened with extinction, within Britain. (After Shirt 1987; Bratton 1991; Hyman and Parsons 1992, 1994.)

	Extinct: formerly native to Britain, but not recorded since 1900	*Threatened*: species listed as 'vulnerable' or 'endangered'[a]	Estimated no. of species surveyed
Insects			
Coleoptera	63	247	3900
Diptera	3	496	6000
Lepidoptera	16	49	2500
(butterflies)	(3)	(5)	(56)
Hymenoptera	18	49	580
Trichoptera	2	13	199
Heteroptera	6	20	540
Orthoptera	–	5	30
Odonata	–	6	41
Total	108	885	14 000
Other invertebrates			
Mollusca	–	17	*c.* 202
Arachnida	–	53	647
Crustacea	1	3	*c.* 70
Myriapoda	–	–	*c.* 92
Others[b]	–	1	*c.* 41
Total	1	74	*c.*1050

[a] For Lepidoptera, includes a few subspecies of macro-moths
[b] Coelenterata, Nemertea, Bryozoa, and Annelida

on the area of the island, A. The islands may be real islands in the sea, or virtual islands, such as freshwater lakes or isolated mountain tops. Very often, log-log plots of S against A show a straight line, so that the relation can be written

$$S = cA^z. \qquad (1)$$

Here, c is a constant, and the parameter z has values in the rough range $z \sim 0.2$–0.3 (MacArthur and Wilson 1967; Diamond and May 1981). This rough rule is often expressed by saying that if the area of suitable habitat is reduced to 10% of its pristine value, the number of species will be halved.

Although less commonly appreciated, the rule also has a theoretical foundation. First, within particular taxonomic groups, the distribution of individuals among species (or 'species-relative abundance') often obeys a so-called 'canonical lognormal distribution' (Preston 1962; MacArthur and Wilson 1967). Such a distribution implies that the total number of individuals, N, is related to the number of species, S, by the power law $S \approx bN^{1/4}$, provided S is reasonably large (May 1975). More generally, a multiplicative interplay among several ecological or evolutionary factors is likely to lead—via the central limit theorem applied to multiplicative random factors—to a lognormal distribution, and for $N \gg 1$ this will tend to give $S \approx bN^z$, with z indeed in the range around 0.2–0.3 and b some constant (May 1975). Second, if we assume that the total number of individuals of the group in question is roughly proportional to the area occupied, we have $N = aA$, with a being a proportionality constant. These rough theoretical arguments thus lead to eq(1), with z in the empirically-observed range.

Suppose we now assume that eq(1) applies to the kinds of species reduction resulting from deforestation or other processes which reduce habitat area. Let the *proportional* reduction in area (expressed as a fraction or as a percentage) be represented by ΔA. If ΔA is small ($\Delta A \ll 1$), the corresponding *proportional* reduction in species numbers, ΔS, is then simply

$$\Delta S = z\, \Delta A. \tag{2}$$

We can now estimate the eventual loss of species consequent upon any particular annual rate of deforestation, ΔA, assuming z indeed is around 0.25. In particular, current rates of tropical deforestation have been variously assigned values in the range of 0.8–2% per year (see, e.g., Reid 1992, table 3.3; and Groom and Schumaker 1993, fig. 2). This immediately translates into a corresponding fraction of 0.2–0.5% of the species in them committed to extinction each year. If sustained, this annual extinction rate is equivalent to an average species' lifespan of around 200–500 years.

To translate any such estimate into a statement about the numbers of species committed to extinction by deforestation obviously requires that we specify the number of species in the group under consideration. If we apply the estimated loss of 0.2–0.5% per year to a conservatively-estimated global total of 5 million species, we arrive at 10–25 thousand species each year, or 1–3 species per hour. This assessment depends not only on our estimates of z and ΔA, but also on our guesstimate of total species numbers; the estimated 200–500 year species' lifespan depends only on z and ΔA.

There are, however, problems with applying such species–area relations to estimated extinction rates. Equation (1) was originally deduced from observations on species distributions among real or virtual islands. If instead applied simply to numbers of species found in areas of different sizes on a continent, eq(1) often does apply, but with z-values more typically in the range of 0.1–0.2 (Diamond and May 1981). As eq(2) makes clear, such a low z-value would imply a lower rate of species' loss for any given rate of loss of habitat area. Probably more important are questions about the extent to which the effects of habitat fragmentation, especially in the tropics, are described by the species–area relations of 'island biogeography'. Simberloff (1992), for example, points out that the forests of the eastern United States were reduced, over two centuries, to fragments totalling only 1–2% of their original extent, yet only three forest birds went extinct.

More generally, the species' lifespans of 200–500 years projected by the species–area methods are much shorter than the lifespans of around 10 000 years deduced earlier for birds and mammals, on the basis of documented extinctions over the past century. Setting aside the fact that these documented extinctions are surely underestimates (so that true lifespans are less than 10 000 years), the discrepancy largely disappears if we are more careful in recognizing that species–area estimates are projected numbers of species that, on current trends of habitat destruction, are 'committed to extinction'. Loss of, say, 90% of habitat will, on this basis, eventually lead to loss of around half the species in it, but the time taken to attain this new, impoverished state is not specified, and will depend on details that vary from place to place and species to species. In general, the notion of 'committed to extinction' recognizes that extinction is a gradual process on an uncertain timescale, but that the endpoint is nevertheless certain once a species' demographic and genetic base has been sufficiently eroded by habitat destruction or other processes (Heywood and Stuart 1992).

For example, Simberloff (1986b) used a species–area approach to estimate that some 1350 species of birds would be committed to extinction by the year 2015. If this figure were misinterpreted as the predicted number actually to go extinct between 1986 and 2015 (which was not Simberloff's intent), it would imply an average of 45 extinctions per year, which seems implausibly high. A more empirical approach is to examine the 1029 entries in Collar and Andrew's (1988) list of threatened bird species. Interpreting 'committed to extinction' as referring to any species whose populations in the wild are no longer viable and will inevitably become extinct, unless major conservation actions reverse current trends (by habitat restoration, elimination of introduced predators, captive breeding and re-introduction, and so on), Heywood and Stuart (1992) estimate that 450 bird species will be committed to extinction by 2015, with 27 of these already probably extinct (in addition to the 12 listed as almost certainly extinct). Whilst this empirically-based estimate is still one-third the theoretically-based one, the disparity is not the worrying factor of 20–50 noted above.

1.7   Estimated future extinction rates from IUCN Red Lists

An alternative approach to estimating impending extinction rates is to look, not at the entries in the IUCN/WCMC database as such, but rather at their recent patterns of change. Specifically, Smith et al. (1993a) have analysed the changes in the IUCN Red Lists for animals in 1986, 1988, and 1990, and in the WCMC Plants Database between 1990 and 1992. The numbers of animal species listed as threatened have increased by more than 30% between 1986 and 1990. Over this interval, only 15 vertebrate species (and 33 animal species in total) satisfied the stringent criteria for addition to the list of recorded extinctions; this rate, if sustained, would take about 7000 years to extinguish half the 47 000 or so vertebrate species. Similarly, between 1990 and 1992, 163 plant species were added to the database of recorded extinctions; this corresponds to about 3000 years for extinction of half the planet's quarter million or so plant species.

We can, however, make a better—although still extremely crude—guess at how extinction rates might currently be accelerating, by examining how fast the status of species on these lists is changing. Smith *et al.* (1993*a*) take the conventional categories of the IUCN lists, and assign them values of 0 for the bottom rung of the ladder (rare, status unknown, or not threatened), 1 for 'vulnerable', 2 for 'endangered', 3 for 'probably extinct', and 4 for 'extinct'. Any species whose status has changed during the time-frame of the analysis is given a score, according to the size and direction of its movement on this ladder. Thus, a change from 'vulnerable' to being 'probably extinct' would score $+2$. The median score for all species in a taxon is then determined; this is a very small number, because the majority of species, even in well-studied groups, are not on any such lists (i.e., have status unchanging at 0). The median time to extinction for species in the group is then computed by dividing the number of status changes required for extinction, starting from unthreatened (viz, $+4$), by the median status change per species per year. For insects, most plants, and other poorly-studied groups, rates of status change tell us more about lack of information and rates of data entry than about extinction. But for birds and mammals, and for the comparatively well-studied palms, they possibly provide a crude guide. For birds and mammals, this approach implies extinction of half the species within roughly 200–300 years, and for palms it suggests 50–100 years.

In Chapter 13, Mace outlines a third, independent way of estimating impending rates of extinction. This again is based on the IUCN categories of threat. However, it uses recent refinements whereby, for some better-known species, these categories are based on species-by-species assessments of extinction probability distributions as functions of time (Mace 1994; Seal *et al.* 1993). Mace calculates the expected times when half the species in each of 10 vertebrate taxa will probably be extinct (3, 4, 3 orders or families of reptiles, birds, mammals, respectively). These 10 average extinction times range from 100 to 1000 years, but are characteristically 300–400 years for mammals and birds.

In this section, we have been estimating the rough times for extinction of half the species in particular taxonomic groups. Given the highly approximate nature of all such estimates, it makes sense to regard them as equivalent to estimates of the projected average lifespan of individual species, until extinction (more strictly, average species' lifespans would differ from the half-life of the entire group by a factor $\ln 2 \simeq 0.7$ if individual extinction was a Poisson process; for a more general discussion of these niceties, see Raup 1978).

It must be emphasized that the data in the Red Lists, upon which both the estimates of Smith *et al.* (1993*a*) and of Mace (Chapter 13) are based, have been compiled opportunistically rather than systematically. They may tell us more about the vagaries of sampling efforts, of taxonomists' interests, and of data entry than about real changes in species' status. Bibby and Mace further underline and document this point in Chapters 7 and 13, respectively. These IUCN category-based approaches nevertheless have potential merit in suggesting new lines of investigation of the dynamics of species extinction. (For instance, Mawdsley and Stork (1994) have taken advantage of the relatively well-documented state of the

British fauna to make analytic comparisons between insects and birds, both for rates of extinction, and for proportions listed in various categories of threat).

In short, all projected extinction rates are beset by many uncertainties. But for birds and mammals, three different approaches—one based on species–area relations, one on IUCN 'category ladder-climbing', one on IUCN estimates of probability distributions—give roughly concordant answers, with projected species' extinction times of around 200–400 years. The situation is less clear for plant species, and even less clear for invertebrates. Moreover, the genetic erosion of plant and other species is a major concern, yet is usually overlooked in extinction debates.

1.8   Extinction rates and conservation planning

The information about extinction processes presented in this book has many implications for conservation planning. In this concluding section, we focus on a few main points.

In the mass extinctions seen in the fossil record, species with limited ranges have, on the whole, tended to suffer most. So it is perhaps not surprising that so many of the threatened species on the IUCN/WCMC lists have very limited ranges. As Bibby discusses more fully in Chapter 7, this suggests we should give priority to identifying and protecting 'hotspots' of endemism, where disproportionate concentrations of threatened plants and animals with restricted ranges are found. Bibby and the International Council for Bird Preservation (ICBP 1992) have noted that 27% of the world's bird species have breeding ranges of less than 50 000 km^2, and that these restricted-range species account for 77% of all threatened birds. They identify an efficient set of 221 Endemic Bird Areas—of which 168, or 76%, are in the tropics—which cover about 5% of the Earth's land area, and yet accommodate 95% of all restricted-range bird species (including 90% of threatened such birds, or 70% of all threatened bird species). This ICBP analysis was prompted, in part, by Myers' (1988) identification of 18 plant 'hotspots', mainly in tropical forests, which in total contain around 20% of all vascular plant species within 0.5% of the world's land surface.

Underlying such focus on particular groups of restricted-range species is the thought that 'hotspots' for, say, birds will also generally be important for plants and other animals (ICBP 1992, p.2). But Thomas and Morris in Chapter 8, for example, show that the factors which endanger British insects vary from group to group, and from habitat to habitat (see also Mawdsley and Stork 1994). In different forms, this theme recurs throughout the book. Particularly telling is the recent analysis by Prendergast *et al.* (1993) of the distribution patterns of five groups—birds, butterflies, dragonflies, aquatic plants, and liverworts—in Britain. Using the detailed (10 km^2 grids) range maps that are available for so many groups in Britain, they found that the 'hotspots' of species-richness for these five different groups did not overlap significantly more than would be expected by chance alone. Even more strikingly, Prendergast *et al.* (1993) showed that, for any one group, the majority of rare species were not to be found in the species-rich places; in these five cases, protecting 'hotspots' would not protect rare species.

In short, endangerment and protection—like commonness and rarity (Rabinowitz *et al.* 1986)—are multidimensional concepts.

Ehrlich's concluding Chapter 14 looks toward a sustainable future, provided we can make monumental changes in the way people live. But even in Ehrlich's most optimistic scenario, we will have enormous impacts on the environment, and upon plant and animal communities, before the rising tide of human numbers levels out and perhaps begins to ebb. Much of the diversity we inherited will be gone before humanity sorts itself out. Under these circumstances, it will be increasingly necessary to make choices about how to optimize what we can save.

In Chapter 12, Margules and Austin show how well-designed surveys and appropriate analysis can provide accurate estimates of the spatial distribution patterns of species. Such databases then allow us to make an efficient choice among potential reserves or protected areas, so as to optimize biological diversity, subject to any given set of constraints (such as total area, and so on). Pressey *et al.* (1993) outline some of the criteria upon which such choices may be based. The simplest such criterion is to preserve the largest number of species from variously-defined pools of candidates.

As Vane-Wright *et al.* (1991) first emphasized, however, not all species are equal. The tuatara, for example, is a large, iguana-like reptile which is the sole survivor of a group that flourished in the Triassic. Today it survives as two species on a few islets off the coast of New Zealand. The tuatara branched off from the main stem of the reptiles' phylogenetic tree so long ago, and is so distinctive, that it comes close to being a two-species sub-class of its own (Daugherty *et al.* 1990). How do we value the tuatara against any other species of reptile? At one democratic extreme, we could regard all species as equally important, each a unique evolutionary product; in this view, the tuatara is no more important than any other among the roughly 6000 species of reptiles. At the opposite extreme, we might give equal weight to each 'sister group' in the phylogenetic tree of reptiles; on this basis, the two species of tuatara would be weighed equally with the sum of all 6000 other reptile species. Vane-Wright *et al.* (1991) propose a sensible middle way, based on the topology of the phylogenetic branching diagram, which seeks to value species according to some rough measure of their evolutionary uniqueness, and which gives results intermediate between the two extremes just outlined (the tuatara, on this scheme, would represent something like a few per cent of the taxonomic distinctness found among reptiles, intermediate between the 0.03% of the democratic extreme and the 50% of the opposite extreme; May 1990*b*).

Various refinements of these basic ideas are being actively pursued (Faith 1992, 1993; Williams *et al.* 1991; Crozier 1992). Ideally, if we had some quantitative measure of the branch lengths within the phylogenetic tree of the group in question, we could unambiguously quantify the amount of 'independent evolutionary history' (IEH) vested in a species, by adding up the lengths of the branches which connect it to the base of the tree and appropriately discounting all shared branches (Faith 1992, 1993; May and Nee in prep.). If we could preserve only, say, half the species in the group, the optimum choice would then be found by maximizing the summed branch length that was preserved. But, generally, we have only the topology of the tree, without quantitative measures of the various branch lengths; in this case, the best procedure would be to assign the branches the lengths that are, on average, most likely for this particular topology, and then go forward on this basis. Such a procedure will, of course, often in fact be sub-optimal, because the underlying evolutionary tree differs from the statistically 'expected' one. In general, however, extensive theoretical simulations of choices made on a topological basis, from artificially generated trees whose underlying branch lengths are known, suggest that values assigned in this way are close to the 'true' ones (May and Nee in prep.). Ultimately, our question is how much of the IEH within a group will be preserved if we can only save, say, 10 of 20 species? The simulations referred to above suggest that, for the 10 of 20 case, we can on average preserve 82% of the group's IEH if we have quantitative information about branch lengths, 77% if we have only topological information about the branching structure of the phylogenetic tree, and 63% if we must choose at random (May and Nee in prep.). Real situations will obviously involve many other important considerations, including other measures of the relative values of species (in preserving 'ecosystem services', for example), and political and economic constraints on which areas may be preserved. But there is no doubt that, increasingly, agonizing choices will have to be made. More and more, such choices will be best made if they can be based on a 'calculus of biodiversity', along the lines just sketched.

Some may ask why be so fussed about impending extinctions. After all, there have been great waves of extinction in the past, and yet diversity has recovered or even rebounded to greater heights. Others will take little comfort from the previous spans of some 10–100 million years for such recovery. Whatever view one takes, the impending sixth mass extinction will be unique in the history of the planet, being the first to result not from environmental changes as such, but rather from the extraordinary population growth and associated activities of one single species. As Ehrlich emphasizes in Chapter 14, *Homo sapiens* now sequesters for its own use somewhere between 25% and 50% of all terrestrial net primary productivity. This state of affairs is without precedent, and in this sense makes the coming extinction spasm qualitatively different from all previous ones.

Summary

The average lifespan of animal species in the fossil record, from origination to extinction, is around 10^6–10^7 years (with the higher number being more typical). For the comparatively well-studied birds and mammals, rates of documented extinction over the past century correspond to species' lifespans of around 10^4 years. And three altogether different methods for projecting impending extinctions—each one of which has serious shortcomings—concur in suggesting a lifespan for bird and mammal species around 200–400 years, if current trends continue. These numbers are likely to be broadly representative of plants and other groups of animals; impending extinction rates are at least 4 orders-of-magnitude faster than the background rates seen in the fossil record.

Such rough estimates are a useful point of departure. As the subsequent chapters make clear, however, what is now needed is a more richly textured understanding of how observed extinction rates differ among taxonomic groups and among geographical locations, and of the underlying causes. Effective conservation action depends upon such understanding.

References

Awramik, S. M., Schopf, J. W. and Walter, M. R. (1983). Filamentous fossil bacteria from the Archean of Western Australia. *Precamb. Res.*, **20**, 357–74.

Bambach, R. K. and Sepkoski, J. J. (1992). The fossil record and marine diversity at different taxonomic levels. *Paleont. Soc. Spec. Publ.*, **6**, 16–36.

Barnes, R. D. (1989). Diversity of organisms: how much do we know? *Amer. Zool.*, **29**, 1075–84.

Benton, M. J. (1985). Patterns in the diversification of Mesozoic non-marine tetrapods and problems in historical diversity analysis. *Paleontology*, **33**, 185–202.

Benton, M. J. (1986). Evolutionary significance of mass extinctions. *TREE*, **1**, 127–30.

Benton, M. J. (1990). Reptiles. In *Evolutionary trends*, (ed. K. J. McNamara). pp. 279–300. Belhaven, London.

Bratton, J. H. (ed.) (1991). *British Red Data Books*, 3. *Invertebrates other than insects.* Joint Nature Conservation Committee, Peterborough.

Busby, J. R. (1986). A biogeoclimatic analysis of *Nothofagus cuninghamii* in southeastern Australia. *Aust. J. Ecol.*, **11**, 1–7.

Collar, N. J. and Andrew, P. (1988). *Birds to watch: The ICPB world check-list of threatened birds*, Technical Publication No. 8. International Council for Bird Preservation, Cambridge.

Crozier, R. H. (1992). Genetic diversity and the agony of choice. *Biol. Conserv.*, **61**, 11–15.

Daugherty, C. H., A. Cree, J. M. Hay, and Thompson, M. B. (1990). Neglected taxonomy and continuing extinctions of tuatara. *Nature*, **47**, 177–9.

Diamond, J. M. (1984). Historic extinctions: a Rosetta stone for understanding prehistoric extinctions. In *Quaternary extinctions: A prehistoric evolution*, (ed. P. S. Martin and R. G. Klein), 824–862. Arizona University Press, Tucson.

Diamond, J. M. (1989). The present, past and future of human-caused extinctions. *Phil. Trans. Roy. Soc. Lond.*, **B325**, 469–77.

Diamond, J. M. and May, R. M. (1981). Island biogeography and the design of natural reserves. In *Theoretical ecology: Principles and applications*, (ed. R. M. May), (2nd edn), Ch. 10. Blackwell, Oxford.

Durham, J. W. (1970). The fossil record and the origin of the *Deuterostomata*. Proceedings of the North American Paleontology Convention, Chicago 1969, Section H. pp. 1104–32. Univ. Chicago Press.

Faith, D. P. (1992). Conservation evaluation and phylogenetic diversity. *Biol. Conserv.*, **61**, 1–10.

Faith, D. P. (1993). Systematics and conservation: on predicting the feature diversity of subsets of taxa. *Cladistics*, **8**, 361–73.

Gaston, K. J. and May, R. M. (1992). The taxonomy of taxonomists. *Nature*, **356**, 281–2.

Grassle, J. F. and Maciolek, N. J. (1992). Deep-sea species richness: regional and local diversity estimates from quantitative bottom samples. *Amer. Nat.*, **139**, 313–41.

Groom, M. J. and Schumaker, N. (1993). Evaluating landscape change: patterns of worldwide deforestation and local fragmentation. In *Biotic interactions and global change*, (ed. P. M. Kareiva, J. G. Kingsolver, and R. B. Huey), pp. 24–44. Sinauer, Sunderland, MA.

Hammond, P. M. (1992). Species inventory. In *Global biodiversity: Status of the Earth's living resources*, (World Conservation Monitoring Centre), pp. 17–39. Chapman & Hall, London.

Hawksworth, D. L. (1991). The fungal dimension of biodiversity: magnitude, significance, and conservation. *Mycol. Res.*, **95**, 441–56.

Heywood, V. H. and Stuart, S. N. (1992). Species extinction in tropical forests. In *Tropical deforestation and species extinction*, (ed. T. C. Whitmore and J. A. Sayer), pp. 91–117. Chapman & Hall, London.

Hyman, P. S. and Parsons, M. S. (ed.) (1992). *A review of the scarce and threatened Coleoptera of Great Britain, Part* 1. Joint Nature Conservation Committee, Peterborough.

Hyman, P. S. and Parsons, M. S. (ed.) (1994). *A review of the scarce and threatened Coleoptera of Great Britain, Part* 2. Joint Nature Conservation Committee, Peterborough.

ICBP (International Council for Bird Preservation) (1992). *Putting biodiversity on the map: Priority areas for global conservation*. ICBP, Cambridge.

IUCN (International Union for the Conservation of Nature and Natural Resources) (1990). 1990 *IUCN Red List of threatened animals*. IUCN, Gland, Switzerland.

Kennedy, W. J. (1977). Ammonite evolution. In *Patterns of evolution*, (ed. A. Hallam), pp. 251–304. Elsevier, Amsterdam.

Knoll, A. H. (1986). Patterns of change in plant communities through geological time. In *Community ecology*, (ed. J. M. Diamond and T. J. Case), pp. 125–41. Harper & Row, New York.

Labandeira, C. C. and Sepkoski, J. J. (1993). Insect diversity in the fossil record. *Science*, **261**, 310–15.

MacArthur, R. H. and Wilson, E. O. (1967). *The theory of island biogeography*. Princeton University Press, Princeton, NJ.

Mace, G. M. (1994). An investigation into methods for categorising the conservation status of species. In *Large scale ecology and conservation biology*, (ed. P. J. Edwards, R. M. May, and N. R. Webb), pp. 295–314. Blackwell, Oxford.

Mace, G. M. and Lande, R. (1991). Assessing extinction threats: toward a reevaluation of IUCN threatened species categories. *Conserv. Biol.*, **5**, 148–57.

Martin, P. S. and Klein, R. G. (1984). *Quaternary extinctions: A prehistoric evolution*. Arizona University Press, Tucson.

Martin, R. D. (1993). Primate origins: plugging the gaps. *Nature*, **363**, 223–34.

Mawdsley, N. A. and Stork, N. E. (1994). Species extinctions in insects: ecological and biogeographical considerations. In *Insects in a changing environment*, (ed. R. Harrington and N. E. Stork). Academic Press, New York, in press.

May, R. M. (1975). Patterns of species abundance and diversity. In *Ecology of species and communities*, (ed. M. Cody and J. M. Diamond), pp. 81–120. Harvard University Press, Cambridge, MA.

May, R. M. (1990a). How many species? *Phil. Trans. Roy. Soc. Lond.*, **B330**, 292–304.

May, R. M. (1990b). Taxonomy as destiny. *Nature*, **347**, 129–30.

May, R. M. (1992). Bottoms up for the oceans. *Nature*, **357**, 278–9.

May, R. M. (1994). Biological diversity: differences between land and sea. *Phil. Trans. Roy. Soc. Lond.*, **B343**, 105–11.

May, R. M. and Nee, S. Making conservation choices: towards a calculus of biodiversity. In prep.

Mohsin, A. K. M. and Ambok, M. A. (1983). *Freshwater fishes of Peninsular Malaysia.* University Pertanian Press, Kuala Lumpar, Malaysia.

Myers, N. (1988). Threatened biotas: 'hotspots' in tropical forests. *Environmentalist*, **8**, 187–208.

Nee, S. and May, R. M. (1992). Patch removal favours inferior competitors. *J. Anim. Ecol.*, **61**, 37–40.

Niklas, K. J., Tiffney, B. H., and Knoll, A. H. (1983). Patterns in vascular land plant diversification. *Nature*, **303**, 614–16.

Nix, H. (1986). A biogeographic analysis of Australian elapid snakes. In *Atlas of elapid snakes of Australia*, (ed. R. C. Longmore), pp. 4–15. Australian Government Printing Service, Canberra.

Patterson, C. and Smith, A. B. (1989). Periodicity in extinction: the role of systematics. *Ecology*, **70**, 802–11.

Poore, G. C. B. and Wilson, G. D. F. (1993). Marine species richness. *Nature*, **361**, 597–8.

Prendergast, J. R., Quinn, R. M., Lawton, J. H., Eversham, B. C., and Gibbons, D. W. (1993). Rare species, the coincidence of diversity hotspots and conservation strategies. *Nature*, **365**, 335–7.

Pressey, R. L., Humphries, C. J., Margules, C. R., Vane-Wright, R. I., and Williams, P. H. (1993). Beyond opportunism: key principles for systematic reserve selection. *TREE*, **8**, 124–8.

Preston, F. W. (1962). The canonical distribution of commonness and rarity. *Ecology*, **43**, 185–215; 410–32.

Rabinowitz, D., Cairns, S., and Dillon, T. (1986). Seven forms of rarity. In *Conservation biology*, (ed. M. E. Soulé), pp. 182–204. Sinauer, Sunderland, MA.

Raup, D. M. (1976). Species diversity on the Phanerozoic: a tabulation. *Paleobiology*, **2**, 279–88.

Raup, D. M. (1978). Cohort analysis of generic survivorship. *Paleobiology*, **4**, 1–15.

Raup, D. M. (1986). Biological extinction in Earth history. *Science*, **231**, 1528–33.

Raup, D. M. (1991). A kill curve for Phanerozoic marine species. *Paleobiology*, **17**, 37–48.

Raup, D. M. (1992). Large-body impact and extinction in the Phanerozoic. *Paleobiology*, **18**, 80–88.

Raup, D. M. and Stanley, S. M. (1978). *Principles of paleontology*, (2nd edn). Freeman, San Francisco.

Reid, W. V. (1992). How many species will there be? In *Tropical deforestation and species extinction*, (ed. T. C. Whitmore and J. A. Sayer), pp. 55–73. Chapman & Hall, London.

Rickards, R. B. (1977). Patterns of evolution in the graptolites. In *Patterns of evolution*, (ed. A. Hallam), pp. 333–58. Elsevier, Amsterdam.

Seal, U. S., Foose, T. J., and Ellis-Joseph, S. (1993). Conservation assessment and management plans (CAMPs) and global captive action plans (GCAPs). In *Creative conservation* (ed. G. M. Mace, P. J. Only, and A. T. C. Feistner), pp. 312–25. Chapman & Hall, London.

Selander, R. K. (1985). Protein polymorphism and the genetic structure of natural populations of bacteria. In *Population genetics and molecular evolution*, (ed. T. Ohta and K. Aoki), pp. 85–106. Springer, Berlin.

Sepkoski, J. J., Jr. (1992). Phylogenetic and ecologic patterns in the Phanerozoic history of marine biodiversity. In *Systematics, ecology, and the biodiversity crisis*, (ed. N. Eldridge), pp. 77–100 Columbia University Press, New York.

Sepkoski, J. J., Jr. and Hulver, M. L. (1985). An atlas of Phanerozoic clade diversity diagrams. In *Phanerozoic diversity patterns: Profiles in macroevolution*, (ed. J. W. Valentine). pp. 11–39. Princeton University Press, Princeton, NJ.

Sepkoski, J. J., Jr., Bambach, R. K., Raup, D. M., and Valentine, J. W. (1981). Phanerozoic marine diversity and the fossil record. *Nature*, **293**, 435–7.

Shirt, D. B. (ed.) (1987). *British Red Data Books*, 2. *insects*. Nature Conservancy Council, Peterborough.

Simberloff, D. S. (1986*a*). The proximate causes of evolution. In *Patterns and processes in the history of life*, (ed. D. M. Raup and D. Jablonski), pp. 259–76. Springer, Berlin.

Simberloff, D. (1986*b*). Are we on the verge of a mass extinction in tropical rain forests? In *Dynamics of extinction*, (ed. D. K. Elliott), pp. 165–80. Wiley, New York.

Simberloff, D. (1992). Do species–area curves predict extinction in fragmented forest? In *Tropical deforestation and species extinction*, (ed. T. C. Whitmore and J. A. Sayer), pp. 75–89. Chapman & Hall, London.

Simpson, G. G. (1952). How many species? *Evolution*, **6**, 342–62.

Smith, F. D. M., May, R. M., Pellew, R., Johnson, T. H., and Walker, K. S. (1993*a*). Estimating extinction rates. *Nature*, **364**, 494–6.

Smith, F. D. M., May, R. M., Pellew, R., Johnson, T. H., and Walker, K. S. (1993*b*). How much do we know about the current extinction rate? *TREE*, **8**, 375–8.

Stork, N. E. (1988). Insect diversity: facts, fiction and speculation. *Biol. J. Linn. Soc.*, **35**, 231–7.

Stork, N. E. and Hine, S. J. Patterns of distribution and abundance of described species as deduced from recent taxonomic revision. Under review.

Stork, N. E. and Lyal, C. J. C. (1993). Extinction or 'co-extinction' rates. *Nature*, **366**, 307.

Tilman, D., May, R. M., Lehman, C. L., and Nowak, M. A. Habitat destruction and the extinction debt. *Nature*, **371**, 65–6.

Valentine, J. W. (1970). How many marine invertebrate fossil species? *J. Paleontol.*, **44**, 410–15.

Vane-Wright, R. I., Humphries, C. J., and Williams, P. H. (1991). What to protect: systematics and the agony of choice. *Biol. Conserv.*, **55**, 235–54.

Van Valen, L. (1973). A new evolutionary law. *Evol. Theory*, **1**, 1–30.

Vermeij, G. J. (1993). Biogeography of recently extinct marine species: implications for conservation biology. *Conserv. Biol.*, **7**, 391–7.

Walter, M. R., Du, R., and Horodyski, R. J. (1990). Coiled carbonaceous megafossils from the Middle Proterozoic of Jixian (Tianjin) and Montana. *Amer. J. Sci.*, **290A**, 133–48.

WCMC (World Conservation Monitoring Centre) (1992). *Global biodiversity: Status of the Earth's living resources*. Chapman & Hall, London.

Whitmore, T. C. and Sayer, J. A. (ed.) (1992). *Tropical deforestation and species extinction*. Chapman & Hall, London.

Williams, P. H., Humphries, C. J., and Vane-Wright, R. I. (1991). Measuring biodiversity: taxonomic relatedness for conservation priorities. *Aust. J. Syst. Bot.*, **4**, 665–79.

Wilson, E. O. (1992). *The diversity of life*. Harvard University Press, Cambridge, MA.

2

Extinctions in the fossil record

David Jablonski

2.1 Introduction

The October 1993 mailing for The Nature Conservancy (US) opens unequivocally: 'Right now, today, species are becoming extinct at a rate *faster* than any time in the Earth's history—one species per day.' Most of the contributors to this volume assess present-day extinction rates, but I am going to address the palaeontological half of that statement. The fossil record provides an enormous database on extinctions, and I will briefly review the fossil record of extinction, how it is quantified and the potential sources of error, and outline some implications for today's biota. Because of significant differences in scale, resolution, and taxa under study, I will argue that the statement quoted above is unjustified, and that profitable comparisons between ancient and ongoing extinctions are less likely to be based on absolute rates than on relative extinction intensities among taxa or regions, on relative recovery rates, or on the biogeographic and evolutionary behaviour of taxa during past intervals of global climate change.

2.2 Global compilations

Most large-scale analyses of extinction in the fossil record rely on synoptic compilations of geologic ranges for genera and families. The fossil record at the species level is very incomplete and especially subject to sampling and preservational biases, and so more inclusive taxa are nearly always used for work at the global scale, particularly over large blocks of geologic time (see Raup 1979a). These synoptic time series are most useful for recognizing episodes of unusual extinction intensity, and for quantifying overall biotic losses. They are less effective in resolving the details of timing or geography for any given extinction event, a research problem requiring an intricate interplay between detailed, local time series, and global compilations (not discussed here; see Raup 1989; Koch 1991; Marshall 1991; Sepkoski and Koch 1994).

As many authors have noted, the fossil record of shelly marine invertebrates is the most complete and reliable for global biodiversity analysis (Benton 1989; Signor 1990 for reviews). Broadly speaking, this is because marine environments provide for more continuous and extensive sedimentation than the terrestrial environment, and shelled invertebrates are both extremely abundant and

Table 2.1 Extinction intensities (percentage extinction, $E/D \times 100$) at the five major mass extinctions in the fossil record. Genus- and family-level values based on Sepkoski's compendia (Sepkoski 1994), with binomial standard errors calculated following Raup (1992a); species-level estimates based on Raup's (1979b) reverse rarefaction technique. Note the agreement between the two estimates of species losses for each extinction episode

	Families		Genera	
Mass extinction	Observed extinction (%)	Calculated species loss (%)	Observed extinction (%)	Calculated species loss (%)
End Ordovician (439 mya)	26 ± 1.9	84 ± 7	60 ± 4.4	85 ± 3
Late Devonian (367 mya)	22 ± 1.7	79 ± 9	57 ± 3.3	83 ± 4
End Permian (245 mya)	51 ± 2.3	95 ± 2	82 ± 3.8	95 ± 2
End Triassic (208 mya)	22 ± 2.2	79 ± 9	53 ± 4.4	80 ± 4
End Cretaceous (65 mya)	16 ± 1.5	70 ± 13	47 ± 4.1	76 ± 5

mya, million years ago; ages from Harland et al. (1990)

intensively studied for economic purposes. Sepkoski's (1992a) *A compendium of marine animal families* has been a boon to all interested in such analyses. Datasets that focus on particular groups, regions or time intervals can be more accurate and more precise, and thus useful for a different and perhaps broader range of questions, but they inevitably lack the sample sizes and temporal sweep of Sepkoski's compendium. Similar databases exist for terrestrial vertebrates and insects (Benton 1989; Maxwell and Benton 1990; Labandiera and Sepkoski 1993); Niklas et al. (1985) analysed the plant record at the species level but have not yet published their data. The terrestrial vertebrates and plants exhibit perturbations that correlate, at least roughly, with the major mass extinctions detected in the oceans, but the magnitude and timing of land-based extinctions relative to the marine ones is still controversial (see McGhee 1989; Johnson and Hickey 1990; Sweet and Braman 1992; Benton 1991; Maxwell 1992; Weems 1992; Erwin 1993).

Synoptic compendia have been the basis of a wide array of intriguing analyses, but here I concentrate on the five major mass extinctions detected for marine invertebrates (Table 2.1). Mass extinctions can be taken as substantial biodiversity losses that are global in extent, taxonomically broad, and rapid relative to the average duration of the taxa involved (Jablonski 1986a). The 'Big Five' extinctions have been confirmed in several generations of family- and genus-level databases (Newell 1952; Valentine 1969; Sepkoski 1994) (Fig. 2.1). The past decade of particularly intensive work, with many additions, corrections, and reinterpretations of taxonomy and stratigraphy have served only to sharpen

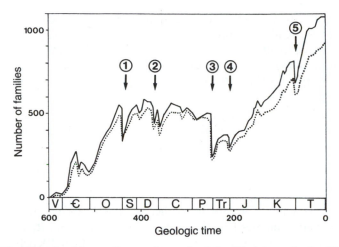

Fig. 2.1 The diversity history for marine animal families through the Phanerozoic, including the 'Big Five' mass extinctions. Solid line, 1992 data; dashed line, 1982 data. Major extinction events: 1, end Ordovician; 2, late Devonian; 3, end Permian; 4, end Triassic; 3, end Cretaceous. Geologic periods along the horizontal axis: V, Vendian, Ꞓ, Cambrian; O, Ordovician; S, Silurian; D, Devonian; C, Carboniferous; P, Permian; Tr, Triassic; J, Jurassic; K, Cretaceous; T, Tertiary. (After Sepkoski 1993.)

these events as seen both in Sepkoski's revised database (Sepkoski 1993) and in a semi-independent compendium by Benton (1993, marshalling an army of taxonomic experts). A host of lesser extinction events are also known; these form local maxima in extinction time series but are not necessarily of the requisite scale geographically (e.g., the early Toarcian: Hallam 1987) or taxonomically (e.g., the late Pleistocene megafauna: Barnosky 1989; Stewart 1991). Some of these local maxima, however, such as the mid Carboniferous extinction pulse, do appear to be genuine but less severe global events (Raymond *et al.* 1990; Sepkoski 1994).

Many taxonomic problems remain in the large databases, of course, and phylogenetic analysis will clarify many of the details of clade survivorship and recovery around extinction events. The inevitable inclusion of paraphyletic taxa, not defined according to strict cladistic principles, is sometimes seen as seriously compromising the entire enterprise (see Smith and Patterson 1988). However, paraphyletic taxa are probably not a major liability at this scale of analysis. Some argue this on biological grounds: the loss of a paraphyletic genus or family often reflects the disappearance of a distinctive mode of life or an ecologically meaningful suite of traits (Van Valen 1984, 1985; Valentine 1990). Sepkoski (1987a, 1989, 1992b) and also Fisher (1991) argue from the topology of evolutionary trees: except for true gradualistic transformations of monospecific taxa, the loss of a paraphyletic genus or family must signal the extinction of at least one species and thus is a useful, if damped, proxy for species-level biodiversity dynamics (Sepkoski and Kendrick 1993 reinforce this point in an elegant simulation study). Similarly, the use of higher taxa in Linnaean rank-based classifications may seem dangerously arbitrary (Smith and Patterson 1988; Doyle and Donoghue 1993). However, under realistic sampling conditions for most groups, even arbitrary higher taxa may provide a more robust portrait of biodiversity changes than will any tabulation of the relatively short-lived and undersampled species themselves (Valentine 1974; Raup 1979a; Gilinsky 1991; Sepkoski and Kendrick 1993). Thus, although a fully resolved phylogenetic analysis is essential for some kinds of problems, such as the diversity histories of specific groups, it is not an absolute prerequisite for reconstructing the broad history of global biodiversity or recognizing the major biotic upheavals in Earth history.

2.3   Extinction metrics

Quantifying extinction intensity has other pitfalls, however, and each of the standard extinction metrics has drawbacks (see Sepkoski and Raup 1986; Raup and Boyajian 1988; Raup 1991a; Gilinsky 1991; Sepkoski and Koch 1994; Foote 1994; for discussions; and Van Valen 1984; Gilinsky and Good 1991, for additional metrics). For example, the raw number of extinctions in a time interval, E, has a simple error term but does not take into account the number of taxa at risk: the loss of 100 families is clearly a more significant event when global diversity (D) totals 200 rather than 600 families. The proportion of taxa becoming extinct (E/D) is intuitively a more satisfying metric but introduces additional uncertainty because estimates of global diversity also have an associated error (Gilinsky 1991).

Extinction rates (E/t) have often been used to take into account the possibility that longer intervals can accumulate more extinctions, but distortions enter owing to uncertainties in geologic time estimates (Sepkoski and Raup 1986; Raup and Boyajian 1988; Gilinsky 1991). For example, the Coniacian, the shortest stage of the Cretaceous, can exhibit an extinction peak in time series analyses (e.g., Sepkoski and Raup 1986; Hubbard and Gilinsky 1992) depending on the timescale used, but exhibits unexceptional turnover under detailed study. More generally, Sepkoski and Koch (1994) show that recent estimates for the duration of stratigraphic stages in the Devonian differ by 26–33% of mean stage length; this uncertainty in the denominator yields alarmingly large error terms. In a simulation study, Foote (1994) found that rate metrics tend to be negatively correlated with interval length under the majority of realistic extinction models. Further, rate metrics assume that extinctions are evenly or randomly distributed through the time interval, which seems not to be true for several of the stratigraphic stages associated with mass extinctions (Raup 1986; Ward 1990; Gilinsky 1991; Foote 1994). Per taxon extinction rates ($E/D/t$) may have the strongest theoretical appeal (Gilinsky 1991), but they compound these uncertainties and thus yield even larger potential error.

The statistical properties of the extinction metrics are still poorly understood, not least because taxonomic databases are not strictly random samples of either diversity or extinctions. Extinction intervals, for example, are often poorly sampled owing to sea-level drops or other environmental changes that reduce the volume of fossiliferous rocks available for study, and rare taxa are themselves under-sampled

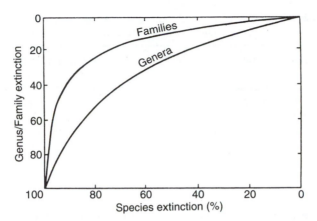

Fig. 2.2 The magnitudes of species-level losses during mass extinctions are estimated using Raup's method of reverse rarefaction. The curves are used to estimate the proportion of species extinction required to account for observed extinctions at the genus or family level. (After Raup 1979**b**.)

but may be more extinction-prone than common taxa. Two principal strategies have been used to minimize these problems: (1) use several metrics to test for sensitivity of interesting features to different error sources (e.g., Van Valen 1984, 1985; Sepkoski and Raup 1986); or (2) use stratigraphic intervals of approximately equal length, combining short intervals and subdividing long ones to reduce variance (e.g., Sepkoski 1989; Raup and Boyajian 1988; Sepkoski and Koch 1994). Confidence intervals are calculated as binomial standard errors. These assume random sampling and thus are inexact, but are useful as measures of relative uncertainty (Raup 1991*a*; Sepkoski and Koch 1994).

2.4 Estimating species-level extinction

Extinction intensities at the species level are estimated from the distribution of species within present-day families and genera (Raup 1979*b*). This 'reverse rarefaction' technique (Fig. 2.2) carries several assumptions. First is the assumption that extinctions are temporally concentrated and not spread through a geologic stage or longer interval. As already noted, some evidence supports this view, but the effects of relaxing this assumption have not been explored systematically. Raup's (1979*b*) estimates were based on the distribution of species within genera and families of present-day echinoids, under the assumption that these distributions were representative of the biota as a whole; this need not be true, particularly in Palaeozoic faunas (see Valentine 1969, 1974; Van Valen 1985). This issue deserves more investigation, but convergent estimates of species losses from both generic or familial data suggests that the echinoid data do provide a reasonably good proxy (see Table 2.1).

Reverse rarefaction also assumes that species survival is uncorrelated with

Table 2.2 Extinction events and taxa in which species-richness was not a buffer against extinction

Cambrian trilobites	Westrop (1989)
End Ordovician bryozoans	Anstey (1978)
Late Devonian corals	Sorauf & Pedder (1986)
Late Devonian ammonoids	House (1985)
End Permian brachiopods	Carlson (1991)
End Cretaceous echinoids	McKinney (1988)
End Cretaceous bivalves and gastropods	Jablonski (1986*b*, 1989)
Exception: End Permian gastropods	Erwin (1989, 1993)

membership in a particular higher taxon. However, related species tend to be clumped ecologically, and in some instances at least (e.g., reef dwellers), their shared risk probably exceeds the random expectation. Species losses would then be overestimated, because it would take fewer non-random species extinctions to remove a given number of genera (see Simberloff 1986, p. 170; Raup 1991*b*, p. 73). The assumption that species loss is randomly distributed among higher taxa also implies that species-poor genera and families should be at greater risk than species-rich ones, which was evidently not the case for many mass extinctions (Table 2.2; but see Erwin 1993 for an exception). This implies that the loss of a genus or family is more likely to involve a species-rich taxon than the random expectation, so that reverse rarefaction might then underestimate species loss. A conservative interpretation of the agreement in Table 2.1, therefore, might be that *genera* are lost randomly with respect to family membership, so that extrapolation from observed extinction at both taxonomic levels yields comparable species-level values.

How far reverse rarefaction distorts true species extinction intensities is unknown. They are probably not severely inaccurate: most studies show species richness to be ineffective rather than actively selected against during mass extinctions, for example. The genus-level extinction data might provide an empirical lower bound for percentage species loss for each event; the loss of a greater proportion of genera than species is unlikely under realistic conditions (D. M. Raup, pers. comm. 18 November 1993). Unusual circumstances could confound such an estimate, however: imagine losing 9 monospecific genera but allow one 10-species genus to survive unscathed, and genus extinction is 90% but species extinction is 47%.

2.5 Comparisons to present-day extinction

Palaeontological extinction data are extremely difficult to compare to present-day extinction rates. As already noted, the most robust palaeontological data, and most analyses on ancient extinction intensities and selectivities, deal with marine invertebrates. The conservation status of today's marine biodiversity is woefully under-documented compared even to our incomplete knowledge of most terrestrial organisms, and only a handful of extinctions are known—

reflecting perhaps both our ignorance and the greater inertia of marine systems (Winston 1992). Human impacts are increasingly severe, however, from over-exploitation to pollution to introduction of exotics, and clearly the marine fossil record can provide useful insights for conservation. The larger problem, however, is how to generalize from the marine-biased history of global biodiversity to today's situation, where most data and the most rapid and extensive species losses are terrestrial.

Even in the marine record, fossil species and higher taxa are not random samples of the biota, or even of the biota with durable skeletons. The taxa in palaeontological databases—marine and terrestrial alike—are skewed towards the more abundant, widespread, and geologically long-lived species, which will have the greatest total number of individuals and occur in the greatest number of localities and rock types, and so are most likely to be preserved and recorded (Raup 1979a; Koch 1991). Moreover, the best time resolution consistently achievable in the pre-Pleistocene fossil record is c. 10^3–10^4 years, due to gaps in the record and time-averaging of successive populations (Kidwell and Behrensmeyer 1993). This does not mean that the *relative* abundances, geographic ranges, and geologic durations of fossil taxa are irretrievable or hopelessly distorted—quantitative analyses of sampling densities and preservation biases have verified that a strong biological signal comes through in many instances (e.g., Paul 1989; Jablonski and Valentine 1990; Marshall 1991; Kidwell and Bosence 1991). But the very nature of the fossil record suggests that palaeontological estimates should be applied to present-day situations with extreme caution, because palaeontological extinction data almost exclusively involve taxa drawn from the extinction-resistant tail of the entire biota! Palaeobiologists continue to debate the kinds of perturbations required to eliminate hundreds or thousands of species with subcontinent-scale geographic ranges (see Jablonski 1991 for examples), and it is not at all clear that the majority of tropical species whose losses are inferred by extrapolation of local surveys—mainly rare, extreme endemics—would be palaeontologically detectable (Jablonski 1991).

Present-day extinction is usually expressed as a simple rate (E/t) rather than as a proportion (E/D) (see Myers 1993 and references therein). However, given the enormous differences in sampled or estimated diversities for extant and fossil biotas, comparisons are only feasible after normalization according to the size of the taxon pool. Smith *et al.* (1993a) tabulated the percentage species loss recorded for different animal and plant groups since 1600, and found none greater than 1.3%. Species threatened by extinction in well-known groups, however, constitute a significantly higher proportion, for example, 32% for gymnosperms and 33% for palms. With some assumptions about extinction probabilities of such groups, these data imply 50% extinction within 50–100 years (Smith *et al.* 1993b). This does approach the intensities required to generate a genus-level extinction on the scale of the 'Big Five' mass extinctions. Still uncertain, of course, is whether these numbers are representative of the entire biota, and how they should be scaled against palaeontological data. The next step might be to develop correction factors for more rigorous comparisons to the fossil record, by estimating for each group the proportion of species whose original abundance and distribution was on a scale commensurate with potential fossilization and palaeontological discovery. This can be done directly using taxa with well-studied fossil records; for example > 77% of shelly marine mollusc species off the California coast occur as Pleistocene fossils (Valentine 1989), and comparable figures could be probably obtained for temperate plant and arthropod species.

2.6 Taxon-specific rates

In the fossil record, species or genus durations within higher taxa vary by more than an order of magnitude even during times of background extinction away from the 'Big Five' mass extinctions, so that computation of a grand mean is not very informative (although broad rate differences do exist among major groups; cf. Stanley 1979; Valentine 1990). Not only is the variance enormous, it is strongly right-skewed: most taxa within a major group tend to be geologically short-lived but the frequency distribution usually includes a tail of long-lived taxa (e.g., Stanley 1979). Some of the observed variation is due to sampling and other artefacts, but data on a wide variety of higher taxa show biological factors to be important in determining extinction rates for species and genera in the fossil record (reviews by Stanley 1985, 1990a). For example, molluscan species durations are positively correlated with geographic ranges (Jablonski 1986b, 1987; Marshall 1991; also Jackson et al. 1985 on corals and bryozoans). Stanley (1990b) found body size to be a more important determinant of survivorship in Late Cenozoic bivalves in the northern Pacific. His attribution of the inverse relation between body size and species survival to differences in population size makes intuitive sense and may well be correct, but needs to be tested more directly. Body size is a poor predictor of species abundances (Blackburn et al. 1993), with population densities in marine communities, for example, ranging over 2–4 orders of magnitude for a given body size except at extreme values (e.g., Marquet et al. 1990).

  Whatever the precise causal links or focal level of selection (cf. Jablonski 1987; Williams 1992), the clear differentials in species survivorship detected in many fossil groups are a rich source of empirical data on variations in extinction risk among taxa (from protozoans, e.g., Norris 1991, to mammals, e.g., Van Valkenburgh and Janis 1993). The list of factors reported to affect extinction probabilities in the fossil record will come as no surprise to the biologist: geographic range, niche breadth, mobility in larval or adult stage, abundance, and population growth rates, etc. The fossil record does more than confirm the role of various attributes, however: it permits the ranking of different factors, providing empirical data on what circumstances erase the benefits of broad geographic range in favor of, say, stable resources—and how large and in what direction the differentials will be among related taxa. Under background extinction, for example, larval ecology is a better predictor of gastropod species survivorship than phylogenetic affinity or adult trophic group (Jablonski 1986c) (Fig. 2.3). Regions, habitats, and taxa should be targeted for palaeobiological

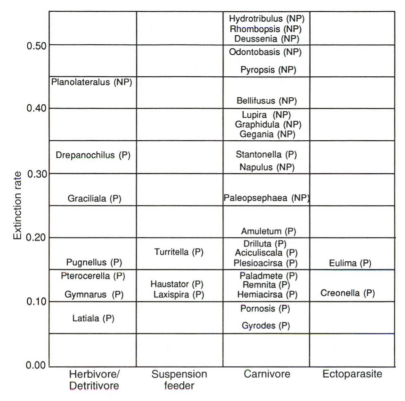

Fig. 2.3 Background extinction rates in Late Cretaceous marine gastropods are more closely related to modes of larval development (P, planktotrophic; NP, non-planktotrophic) than to the adult feeding categories listed along the bottom of the figure. These significant differences in extinction rate are probably the result of the broader larval dispersal ability, and thus broader geographic ranges, typical of species having planktotrophic larvae ($P < 0.001$, Mann–Whitney U-test) (see Jablonski 1986a, and in prep.). Extinction rates calculated as a per-species rate per million years.

studies designed explicitly to address conservation issues; palaeobotanical analyses along these lines are sorely needed, for example.

It may be surprising that I advocate palaeontological analyses of background extinction as a tool in assessing present-day extinction risk, given data suggesting that background rules of survivorship are disrupted during the major mass extinctions (e.g., Jablonski 1986b, 1989, 1991; Westrop 1991) (Table 2.2). I am not belittling the magnitude of today's problems, nor do I deny the potential for long-term losses of similar scope and evolutionary impact to the major mass extinctions of the fossil record (cf. Simberloff 1986). Background extinction patterns may nevertheless provide clues to how the present-day biota will respond to anthropogenic disturbances, for at least three reasons:

1. It is not clear that present-day disturbances, although undeniably extensive,

Table 2.3   Extinction events and taxa in which broad geographic range at the genus level enhanced survivorship

Late Cambrian trilobites	Fortey (1983); Westrop (1989, 1991)
End Ordovician bivalves	Bretsky (1973)
End Ordovician brachiopods	Sheehan & Coorough (1990)
End Ordovician bryozoans	Anstey (1986)
End Ordovician trilobites	Robertson *et al.* (1990)
Late Devonian bivalves	Bretsky (1973)
End Permian bivalves	Bretsky (1973)
End Permian gastropods	Erwin (1989, 1993)
End Triassic bivalves	Bretsky (1973)
End Cretaceous bivalves and gastropods	Jablonski (1986*b*, 1989)

are on a par with those that drove the major mass extinctions. If we could apply all the appropriate correctives, present and near-future extinctions may or may not fall quantitatively within palaeontological background rates but there is little evidence that the qualitative change in survivorship such as seen at the Cretaceous–Tertiary boundary has occurred today. So far as they are known, today's extinction patterns conform mainly to intensified versions of background expectations, with losses concentrated in endemic species and subspecies. The major mass extinctions operated on a different scale: genera endemic to single subcontinental provinces were lost preferentially, regardless of the geographic ranges of their constituent species (Jablonski 1986*a*, 1989, 1991) (Table 2.3).

2. Even if selectivities of present-day extinctions were congruent with those seen during the ancient mass extinctions, not all background patterns of survival are overturned under the mass extinction regime, and some evolutionary continuity exists even across the most severe events (Jablonski 1986*d*, 1989). The persistence of survival advantages may be clade- or event-specific, as in the presence of a resting cyst in the life cycles of high-latitude phytoplankton during the end Cretaceous event (Kitchell *et al.* 1986; see also Norris 1991 on continuity in planktic Foraminifera). In other instances, enhanced survivorship under both mass and background regimes may represent the long-term effects of previous mass extinctions. For example, severe extinction in the bivalve order Pholado-myoida during the end Permian extinction but not during the succeeding end Triassic event may reflect the purging of vulnerable members of the clade followed by re-radiation from survivors that happened to retain extinction-resistant features. Similarly, the remarkably low extinction rates in shallow-water molluscs during the huge Pleistocene climate and sea-level fluctuations (references in Valentine and Jablonski 1993) have been linked to Pliocene turnover events that may have already eliminated the most vulnerable species (Stanley 1990*b*; see also Jackson, Chapter 3). Because we are just beginning to explore the factors that allow clades or trends to prevail under both background and mass extinction regimes, the limits to the relevancy of background extinction

patterns to perturbations of various magnitudes—particularly those short of the 'Big Five' events—are still unclear.

3. The great majority, probably >90%, of species extinctions in the fossil record occur outside the five major extinction events (Raup 1986, 1991c): mass extinctions have such profound biological consequences because they bite deep into standing diversity and disrupt background selection regimes, not because they account for most species terminations. In fact, extinction magnitudes for the stratigraphic stages of the Phanerozoic form a continuous distribution. Some impressive extinction pulses fail to stand significantly above background variance, and data from those intervals should be used to advantage as long-term records of taxon selectivity.

Geographic patterns

For present-day biotas, extinction risk is not spread evenly over the globe. Some regions are suffering greater encroachment than others, and some regions are known to be exceptionally species-rich so that their disruption will be especially costly in terms of biodiversity. The biogeography of ancient mass extinctions has been relatively neglected, perhaps because such data are not reliably recorded in synoptic databases, but a few general patterns emerge.

Most striking is the disruption of reef and related communities on tropical carbonate shelves: each mass extinction brings a major reorganization of these taxon-rich habitats, with once-dominant taxa eliminated or relegated to secondary roles (Copper 1988, 1989; Talent 1988; Kauffman and Fagerstrom 1993). Alternative hypotheses for these spectacular collapses include: (a) tropical biotas in general are fragile because their species are adapted to a narrow range of climatic and other conditions; (b) tropical biotas contain a large proportion of extinction-prone endemics, so that losses are strong here owing to biogeographic structure; (c) reef communities are such a tightly woven network of biological interactions that the initial removal of the same proportion of species as were lost at high latitudes could be more disruptive; (d) the favoured habitat of reef communities, low-sedimentation and low-nutrient shallow-water platforms or ramps, is itself easily disrupted. Some support exists for the biogeography- and habitat-based hypotheses. First, interprovincial variation in mass extinction intensities within latitudinal belts tend to be positively related to the proportion of endemic genera in the pre-extinction biota (Jablonski 1989; Sheehan and Coorough 1990; Westrop 1991). Second, a global analysis of end-Cretaceous extinction in marine bivalves found that tropical settings outside of the carbonate platforms suffered no greater losses than did extratropical faunas (Raup and Jablonski 1993). The dissection of these alternative mechanisms for the repeated demise of reef communities, indeed for any major community type, could have valuable applications to present-day biodiversity.

As discussed above, marine extinctions are remarkably sparse over the past 2 million years, even for reef taxa, despite massive fluctuations in sea-level and other environmental variables as global climate oscillated between glacial and interglacial states. Local extinctions were legion, but few taxa disappeared entirely because species were able to adjust geographic ranges in response to dramatic shifts in climate and shoreline—an option increasingly unavailable, I might add, given human encroachments (review in Valentine and Jablonski 1993). The same phenomenon has been superbly documented for terrestrial animals and plants (see Graham and Grimm 1990; Webb and Bertlein 1992; Coope, Chapter 4) and has immediate implications for the design of biological reserves. Reserves must be sufficiently large and environmentally complex to accommodate the array of disparate geographic range shifts that any climate change will evoke from the resident species (Graham 1988; Hunter et al. 1988). Reserves that are environmentally homogeneous or constructed tightly around the present distribution of a cluster of high-priority species will see their subjects melt away species by species in response to climate changes. This disassembly process has occurred not only in the temperate latitudes that bore the brunt of Pleistocene glaciations, but in tropical settings as well (marine, Paulay 1990; terrestrial, Bush et al. 1992).

2.7 Biodiversity and biodisparity

Other palaeontological approaches can be used to explore the effects of extinction, such as quantifying the loss of morphological variety rather than taxonomic diversity *per se*. The ecological or evolutionary impact of an extinction event resides not simply in the number of taxa lost, but in the loss of what could be termed (with some trepidation) *biodisparity*, the range of morphologies or other attributes within a clade or within a local or regional subsample of a clade. Neither the will nor the resources exist to save every endangered species in today's biosphere, and when priorities are set, biodisparity should probably enter into the equation. Phylogenetic information is also important in many conservation contexts; some workers have proposed metrics for phylogenetic distinctness, for example, based on depth of branch-points in a phylogeny or number of species relative to sister taxa (reviewed by Nixon and Wheeler 1992).

A drawback of phylogenetic metrics is their requirement of detailed cladistic analyses; time is not available for such analyses for all potentially endangered plants and animals. Further, consistency indices decline and the number of alternative or slightly less parsimonious phylogenetic trees expands with the number of taxa analysed, so that subgroups can radically change their conservation status with alternative interpretations of character state distributions, or with a glance at those trees that are one or two steps longer than the shortest phylogeny. Genealogy is extremely important in understanding biodiversity, but it should not be the only basis for weighing biological 'quality' against 'quantity'. If we are concerned with avoiding the loss of particular functional groups, or with maximizing the potential source pool for evolutionary recovery, then biodisparity measures may provide a more appropriate assessment, beyond sheer numbers of taxa, of how priorities should be set.

A variety of methods are available for quantifying biodisparity, ranging from ecomorphological studies that test for convergence or displacement of species morphologies within and among communities (e.g., Winemiller 1991) to the construction of theoretical or empirical morphospaces used to track clades through geologic time (see McGhee 1991 and Foote 1991, 1993). Most of the palaeobiological work has focused on radiations, particularly the relation between taxonomic and morphologic diversification during the Cambrian Explosion, but background and mass extinction could be analysed in the same fashion. At present, we know little of how taxonomic losses impinge on patterns of morphospace occupation in ancient or present-day biotas. As Foote (1992) points out, if extinction is effectively random with respect to morphology, a disproportionately large number of extinctions is required to reduce morphological variety substantially. On the other hand, extinctions might be selective in morphospace just as they are taxonomically. This has not been explored, however, and we do not know if taxa lying in different regions or densities of morphospace vary predictably in risk. For example, are there kinds of perturbations (density-dependent versus density-independent?) that tend to concentrate losses in the core morphospace or around the morphospace periphery? Do different higher taxa or extinction intensities exhibit characteristic loss patterns of morphospace?

Biodisparity analyses should not be seen as a new way to identify morphological outliers that can serve as flagship species for biological preserves. Instead, they can quantify real or potential losses in a region's or community's relative biological wealth in terms of morphologies rather than species, potentially even in the absence of detailed formal taxonomy (cf. Foote 1993).

2.8 Rebounds

Evolutionary rebounds after mass extinctions are an important component of macroevolution (reviews by Jablonski 1986b, c; Benton 1987). The recovery of both biodiversity and biodisparity in the aftermath of mass extinctions is rapid by geological timescales, often accompanied by significantly accelerated evolutionary rates (Hallam 1987; Miller and Sepkoski 1988; Sheehan and Coorough 1990; Sepkoski 1992b), but it is extremely slow by human timescales. Refurbished reef communities, for example, emerge only after a 5–10 million year lag following each of the major mass extinctions (Copper 1988, 1989; Kauffman and Fagerstrom 1993) (Fig. 2.4). Talent (1988) argues that this delay stretches far beyond the offset of the environmental stresses that eliminated the preceding community. This has significant implications for restoration ecology, as it implies some other constraint on the evolution of species or the assembly of communities capable of occupying these habitats; the community ecology of rebounds from past extinction events should be a focus for future research. In addition, recoveries from global events need not be geographically homogeneous. For example, the post-Cretaceous rebounds in molluscan faunas differ significantly in Europe and North America despite the lack of geographic variation in end-Cretaceous extinction intensities (references in Raup and Jablonski 1993).

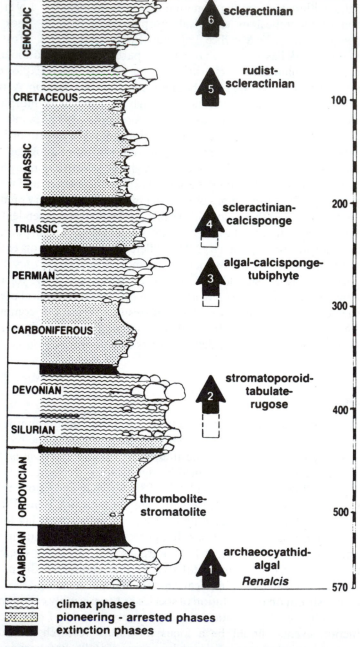

Fig. 2.4 The geologic history of reef and carbonate platform faunas. Six major associations are recognized, all but the present scleractinian-dominated one terminated by an extinction event, followed by a protracted interval lacking reefs. (From Copper 1988, by permission of the author and the Society for Sedimentary Geology.)

2.9 Conclusion

The mass extinctions of the fossil record are, if nothing else, cautionary tales. They show unequivocally that marine and terrestrial biotas are not infinitely resilient, but have breaking points that have been exceeded repeatedly in the past, with extreme and long-lasting biological consequences. The fossil record of extinction can yield more specific insights, however, even though estimates of present-day extinction are difficult to compare directly with those calculated for past mass extinctions. Raw numbers aside, the fossil record provides our only empirical data on what happens when biological communities collapse or disassemble, when increased extinction rates impinge on taxa of different relative vulnerabilities, when global warming or cooling occurs faster than species can adjust to local conditions, when ecological stresses ameliorate after prolonged or severe episodes, and so on. All of these are situations that we may face in the coming century, and it is in these more circumscribed questions that the fossil record may provide the most insight and has the greatest predictive power.

Summary

Direct comparison of ancient extinctions to the present-day situation is difficult, because quantitative palaeontological data come primarily from marine invertebrates, fossilized species are usually drawn from the more abundant and widespread taxa, and time resolution is rarely better than 10^3–10^4 years. A growing array of techniques permits quantitative error estimates on some of these potential biases, and allows calculation of species extinction intensities from genus-level data, which are more robust. Extensive as today's species losses probably are, they have yet to equal any of the 'Big Five' mass extinctions. Background extinction patterns are potential sources of insight regarding present-day biotic losses; over 90% of past species extinction has occurred at times other than the 'Big Five' mass extinctions. Mean durations of fossil species vary by more than an order of magnitude even within clades, rendering uninformative any global average for background extinction. Taxon-specific variation is evidently related to intrinsic biotic factors such as geographic range and population size. Approaches to extinction analysis and prediction based on morphological variety or *biodisparity*, rather than biodiversity, should be explored as an adjunct or alternative to taxon inventories or phylogenetic metrics. Finally, rebounds from mass extinctions are geologically rapid but ecologically slow, and biodiversity recovery and the reestablishment of some communities typically requires 5–10 million years.

Acknowledgements

I thank D. H. Erwin, S. M. Kidwell, J. H. Lawton, R. M. May, D. M. Raup, and J. J. Sepkoski, Jr., for valuable discussions and reviews, and the National Science Foundation for support.

References

Anstey, R. L. (1978). Taxonomic survivorship and morphologic complexity in Paleozoic bryozoan genera. *Paleobiology*, **4**, 407–18.

Anstey, R. L. (1986). Bryozoan provinces and patterns of generic evolution and extinction in the Late Ordovician of North America. *Lethaia*, **19**, 33–51.

Barnosky, A. D. (1989). The late Pleistocene event as a paradigm for widespread mammal extinction. In *Mass extinctions*, (ed. S. K. Donovan), pp. 235–54. Columbia University Press, New York.

Benton, M. J. (1987). Progress and competition in macroevolution. *Biol. Rev.*, **62**, 305–38.

Benton, M. J. (1989). Mass extinctions among tetrapods and the quality of the fossil record. *Phil. Trans. Roy. Soc. Lond.*, **B325**, 369–86.

Benton, M. J. (1991). What really happened in the Late Triassic? *Hist. Biol.*, **5**, 263–78.

Benton, M. J. (ed.) (1993). *The fossil record*, (2nd edn.) Chapman & Hall, London.

Blackburn, T. M., Brown, V. K., Doube, B. M., Greenwood, J. J. D., Lawton, J. H., and Stork, N. E. (1993). The relationship between abundance and body size in natural animal assemblages. *J. Anim. Ecol.*, **62**, 519–28.

Bretsky, P. W. (1973). Evolutionary patterns in the Paleozoic Bivalvia: Documentation and some theoretical considerations. *Geol. Soc. Am. Bull.*, **84**, 2079–96.

Bush, M. B. *et al.* (1992). A 14 300-yr paleoecological profile of a lowland tropical lake in Panama. *Ecol. Monogr.*, **62**, 251–75.

Carlson, S. J. (1991). A phylogenetic perspective on articulate brachiopod diversity and the Permo-Triassic extinctions. In *The unity of evolutionary biology* (ed. E. C. Dudley), pp. 119–42. Dioscorides Press, Portland, Oregon.

Copper, P. (1988). Ecological succession in Phanerozoic reef ecosystems: Is it real? *Palaios*, **4**, 424–38.

Copper, P. (1989). Enigmas in Phanerozoic reef development. *Mem. Assoc. Australas. Palaeontol.*, **8**, 371–85.

Doyle, J. A. and Donoghue, M. J. (1993). Phylogenies and angiosperm diversification. *Paleobiology*, **19**, 141–67.

Erwin, D. W. (1989). Regional paleoecology of Permian gastropod genera, southwestern United States and the end-Permian mass extinction. *Palaios*, **4**, 424–38.

Erwin, D. H. (1993). *The great Paleozoic crisis: Life and death in the Permian*. Columbia University Press, New York.

Fisher, D. C. (1991). Phylogenetic analysis and its application in evolutionary paleobiology. In *Analytical paleontology. Short courses in paleontology*, No. 4, (ed. N. L. Gilinsky and P. W. Signor), pp. 103–23. The Paleontological Society, University of Tennessee, Knoxville.

Foote, M. (1991). Analysis of morphological data. In *Analytical paleontology. Short courses in paleontology*, No. 4, (ed. N. L. Gilinsky and P. W. Signor), pp. 59–86. The Paleontological Society, Knoxville, University of Tennessee.

Foote, M. (1992). Rarefaction analysis of morphological and taxonomic diversity. *Paleobiology*, **18**, 1–16.

Foote, M. (1993). Discordance and concordance between morphological and taxonomic diversity. *Paleobiology*, **19**, 185–204.

Foote, M. (1994). Temporal variation in extinction risk and temporal scaling of extinction metrics. *Paleobiology*, **20**, 00–00.

Fortey, R. A. (1983). Cambrian-Ordovician trilobites from the boundary beds in western Newfoundland and their phylogenetic significance. *Spec. Pap. Palaeontol.*, **30**, 179–211.

Gilinsky, N. L. (1991). The pace of taxonomic evolution. In *Analytical paleontology. Short courses in paleontology*, No. 4, (ed. N. L. Gilinsky and P. W. Signor), pp. 157–74. The Paleontological Society, University of Tennessee, Knoxville.

Gilinsky, N. L. and Good, I. J. (1991). Probabilities of origination, persistence, and extinction of families of marine invertebrate life. *Paleobiology*, **17**, 145–67.

Graham, R. W. (1988). The role of climatic change in the design of biological reserves: The paleoecological perspective for conservation biology. *Conserv. Biol.*, **2**, 391–4.

Graham, R. W. and Grimm, E. C. (1990). Effects of global change on the patterns of terrestrial biological communities. *Trends Ecol. Evol.*, **5**, 289–92.

Hallam, A. (1987). Radiations and extinctions relative to environmental change in the marine Lower Jurassic of northwest Europe. *Paleobiology*, **13**, 152–68.

Harland, W. B., Armstrong, R. L., Cox, A. V., Craig, L. E., Smith, A. G., and Smith, D. G. (1990). *A geologic time scale 1989*. Cambridge University Press.

House, M. R. (1985). Correlation of mid-Palaeozoic ammonoid evolutionary events with global sedimentary perturbations. *Nature*, **313**, 17–22.

Hubbard, A. E. and Gilinsky, N. L. (1992). Mass extinctions as statistical phenomena: An examination of the evidence using χ^2 tests and bootstrapping. *Paleobiology*, **18**, 148–60.

Hunter, M. L., Jr., Jacobson, G. L., Jr., and Webb, T., III (1988). Paleoecology and the coarse-filter approach to maintaining biological diversity. *Conserv. Biol.*, **2**, 375–85.

Jablonski, D. (1986a). Causes and consequences of mass extinctions: A comparative approach. In *Dynamics of extinction*, (ed. D. K. Elliott), pp. 183–229. Wiley, New York.

Jablonski, D. (1986b). Background and mass extinctions: The alternation of macroevolutionary regimes. *Science*, **231**, 129–33.

Jablonski, D. (1986c). Larval ecology and macroevolution in marine invertebrates. *Bull. Mar. Sci.*, **39**, 565–87.

Jablonski, D. (1986d). Evolutionary consequences of mass extinctions. In *Patterns and processes in the history of life*, (ed. D. M. Raup and D. Jablonski), pp. 313–29. Springer, Berlin.

Jablonski, D. (1987). Heritability at the species level: Analysis of geographic ranges of Cretaceous mollusks. *Science*, **238**, 360–3.

Jablonski, D. (1989). The biology of mass extinction: A palaeontological view. *Phil. Trans. Roy. Soc. Lond.*, **B325**, 357–68.

Jablonski, D. (1991). Extinctions: A paleontological perspective. *Science*, **253**, 754–7.

Jablonski, D. and Valentine, J. W. (1990). From regional to total geographic ranges: Testing the relationship in Recent bivalves. *Paleobiology*, **16**, 126–42.

Jackson, J. B. C., Winston, J. E., and Coates, A. G. (1985). Niche breadth, geographic range, and extinction of Caribbean reef-associated cheilostome Bryozoa and Scleractinia. In *Proceedings of the 5th International Coral Reef Congress*, Vol. 4, pp. 151–8.

Johnson, K. R. and Hickey, L. J. (1990). Megafloral change across the Cretaceous/Tertiary boundary in the northern Great Plains and Rocky Mountains, U.S.A. *Geol. Soc. Am. Spec. Paper*, 247, 433–44.

Kauffman, E. G. and Fagerstrom, J. A. (1993). The Phanerozoic evolution of reef diversity. In *Species diversity in ecological communities*, (ed. R. E. Ricklefs and D. Schluter), pp. 315–29. University of Chicago Press.

Kidwell, S. M. and Behrensmeyer, A. K. (ed.) (1993). *Taphonomic approaches to time resolution in fossil assemblages*. *Short courses in paleontology*, No. 6. The Paleontological Society, University of Tennessee, Knoxville.

Kidwell, S. M. and Bosence, D. W. J. (1991). Taphonomy and time-averaging of marine shelly faunas. In *Taphonomy*, (ed. P. A. Allison and D. E. G. Briggs), pp. 116–209. Plenum, New York.

Kitchell, J. A., Clark, D. L., and Gombos, A. M., Jr. (1986). Biological selectivity and extinction: A link between background and mass extinction. *Palaios*, **1**, 504–11.

Koch, C. F. (1991). Sampling from the fossil record. In *Analytical paleontology. Short courses in paleontology*, No. 4, (ed. N. L. Gilinsky and P. W. Signor), pp. 4–18. The Paleontological Society, University of Tennessee, Knoxville.

Labandeira, C. C. and Sepkoski, J. J., Jr. (1993). Insect diversity in the fossil record. *Science*, **261**, 310–15.

Marquet, P. A., Navarrete, S. A., and Castilla, J. C. (1990). Scaling population density to body size in rocky intertidal communities. *Science*, **250**, 1125–7.

Marshall, C. R. (1991). Estimation of taxonomic ranges from the fossil record. In *Analytical paleontology. Short courses in paleontology*, No. 4, (ed. N. L. Gilinsky and P. W. Signor), pp. 19–38. The Paleontological Society, University of Tennessee, Knoxville.

Maxwell, W. D. (1992). Permian and Early Triassic extinction of nonmarine tetrapods. *Palaeontology*, **35**, 571–83.

Maxwell, W. D. and Benton, M. J. (1990). Historical tests of the absolute completeness of the fossil record of tetrapods. *Paleobiology*, **16**, 322–35.

McGhee, G. R., Jr. (1989). The Frasnian-Famennian extinction event. In *Mass extinctions*, (ed. S. K. Donovan), pp. 133–51. Columbia University Press, New York.

McGhee, G. R., Jr. (1991). Theoretical morphology: The concept and its applications. In *Analytical paleontology. Short courses in paleontology*, No. 4, (ed. N. L. Gilinsky and P. W. Signor), pp. 87–102. The Paleontological Society, University of Tennessee, Knoxville.

McKinney, M. L. (1988). Extinction selectivity: A key to macroevolutionary processes [Abstract]. *Geol. Soc. Am. Abstr.*, **20**, A205.

Miller, A. I. and Sepkoski, J. J., Jr. (1988). Modeling bivalve diversification: The effect of interaction on a macroevolutionary system. *Paleobiology*, **14**, 364–9.

Myers, N. (1993). Questions of mass extinction. *Biodiversity and Conservation*, **2**, 2–17.

Newell, N. D. (1952). Periodicity in invertebrate evolution. *J. Paleontol.*, **26**, 371–85.

Niklas, K. J., Tiffney, B. H., and Knoll, A. H. (1985). Patterns in vascular land plant diversification: A factor analysis at the species level. In *Phanerozoic diversity patterns*, (ed. J. W. Valentine), pp. 97–128. Princeton University Press, Princeton, NJ.

Nixon, K. C. and Wheeler, Q. D. (1992). Measures of phylogenetic diversity. In *Extinction and phylogeny*, (ed. M. J. Novacek and Q. D. Wheeler), pp. 216–34. Columbia University Press, New York.

Norris, R. D. (1991). Biased extinction and evolutionary trends. *Paleobiology*, **17**, 388–99.

Paul, C. R. C. (1989). Patterns of evolution and extinction in invertebrates. In *Evolution and the fossil record*, (ed. K. C. Allen and D. E. G. Briggs), pp. 99–121. Smithsonian Institution Press, Washington, DC.

Paulay, G. (1990). Effects of late Cenozoic sea-level fluctuations on the bivalve faunas of tropical oceanic islands. *Paleobiology*, **16**, 415–34.

Raup, D. M. (1979a). Biases in the fossil record of species and genera. *Bull. Carnegie Mus. Nat. Hist.*, **13**, 85–91.

Raup, D. M. (1979b). Size of the Permo-Triassic bottleneck and its evolutionary implications. *Science*, **206**, 217–18.

Raup, D. M. (1986). Biological extinction in Earth history. *Science*, **231**, 1528–33.

Raup, D. M. (1989). The case for extraterrestrial causes of extinction. *Phil. Trans. Roy. Soc. Lond.*, **B325**, 421–35.

Raup, D. M. (1991a). The future of analytical paleontology. In *Analytical paleontology.*

Short courses in paleontology, No. 4, (ed. N. L. Gilinsky and P. W. Signor), pp. 207–16. The Paleontological Society, University of Tennessee, Knoxville.

Raup, D. M. (1991*b*). *Extinction: Bad genes or bad luck?* Norton, New York.

Raup, D. M. (1991*c*). A kill curve for Phanerozoic marine species. *Paleobiology*, **17**, 37–48.

Raup, D. M. and Boyajian, G. E. (1988). Patterns of generic extinction in the fossil record. *Paleobiology*, **14**, 109–25.

Raup, D. M. and Jablonski, D. (1993). Geography of end-Cretaceous marine bivalve extinctions. *Science*, **260**, 971–3.

Raymond, A., Kelley, P. H., and Lutken, C. B. (1990). Dead by degrees: Articulate brachiopods, paleoclimate and the mid-Carboniferous extinction event. *Palaios*, **5**, 111–23.

Robertson, D. B. R., Brenchley, P. J., and Owen, A. W. (1990). Ecological disruption close to the Ordovician-Silurian boundary. *Hist. Biol.*, **5**, 131–44.

Sepkoski, J. J., Jr. (1987*a*). Is the periodicity of extinctions a taxonomic artifact? Reply. *Nature*, **330**, 251–2.

Sepkoski, J. J., Jr. (1989). Periodicity in extinction and the problem of catastrophism in the history of life. *J. Geol. Soc. Lond.*, **146**, 7–19.

Sepkoski, J. J., Jr. (1992*a*). *A compendium of fossil marine animal families*, (2nd edn). Milwaukee Pub. Mus. Contrib. Biol. Geol., Vol. 83.

Sepkoski, J. J., Jr. (1992*b*). Phylogenetic and ecologic patterns in the Phanerozoic history of marine biodiversity. In *Systematics, ecology and the biodiversity crisis*, (ed. N. Eldredge), pp. 77–100. Columbia University Press, New York.

Sepkoski, J. J., Jr. (1993). Ten years in the library: New data confirm paleontological patterns. *Paleobiology*, **19**, 43–51.

Sepkoski, J. J., Jr. (1994). Patterns of Phanerozoic extinction: A perspective from global data bases. In *Global bio-events and event-stratigraphy*, (ed. O. H. Walliser). Springer, Berlin, in press.

Sepkoski, J. J., Jr. and Kendrick, D. C. (1993). Numerical experiments with model monophyletic and paraphyletic taxa. *Paleobiology*, **19**, 168–84.

Sepkoski, J. J., Jr. and Koch, C. F. (1994). Evaluating paleontologic data relating to bio-events. In *Global bio-events and event-stratigraphy*, (ed. O. H. Walliser). Springer, Berlin, in press.

Sepkoski, J. J., Jr. and Raup, D. M. (1986). Periodicity of marine extinction events. In *Dynamics of extinction*, (ed. D. K. Elliott), pp. 3–36. Wiley, New York.

Sheehan, P. M. and Coorough, P. J. (1990). Brachiopod zoogeography across the Ordovician-Silurian boundary. *Geol. Soc. Lond. Mem.*, **12**, 181–7.

Signor, P. W. (1990). The geologic history of diversity. *Annu. Rev. Ecol. Syst.*, **21**, 509–39.

Simberloff, D. (1986). Are we on the verge of a mass extinction in tropical rain forests? In *Dynamics of extinction*, (ed. D. K. Elliott), pp. 165–80. Wiley, New York.

Smith, A. B. and Patterson, C. (1988). The influence of taxonomic method on the perception of patterns of evolution. *Evol. Biol.*, **23**, 127–216.

Smith, F. D. M., May, R. M., Pellew, R., Johnson, T. H., and Walter, K. R. (1993*a*). How much do we know about the current extinction rate? *Trends Ecol. Evol.*, **9**, 375–8.

Smith, F. D. M., May, R. M., Pellew, R., Johnson, T. H., and Walter, K. R. (1993*b*). Estimating extinction rates. *Nature*, **364**, 494–6.

Sorauf, J. E. and Pedder, A. E. H. (1986). Late Devonian rugose corals and the Frasnian-Famennian crisis. *Can. J. Earth Sci.*, **23**, 1265–87.

Stanley, S. M. (1979). *Macroevolution*. W. H. Freeman, San Francisco.

Stanley, S. M. (1985). Rates of evolution. *Paleobiology*, **11**, 13–26.

Stanley, S. M. (1990*a*). The general correlation between rate of speciation and rate of extinction: Fortuitous causal linkages. In *Causes of evolution*, (ed. R. M. Ross and W. D. Allmon), pp. 103–27. University of Chicago Press.

Stanley, S. M. (1990*b*). Delayed recovery and the spacing of major extinctions. *Paleobiology*, **16**, 401–14.

Stewart, A. J. (1991). Mammalian extinctions in the Late Pleistocene of northern Eurasia and North America. *Biol. Rev.*, **66**, 453–562.

Sweet, A. R. and Braman, D. R. (1992). The K/T boundary and contiguous strata in western Canada: Interactions between paleoenvironments and palynological assemblages. *Cret. Res.*, **13**, 31–79.

Talent, J. A. (1988). Organic reef-building: Episodes of extinction and symbiosis? *Senckenb. Lethaea*, **69**, 315–68.

Valentine, J. W. (1969). Patterns of taxonomic and ecologic structure of the shelf benthos during Phanerozoic time. *Palaeontology*, **12**, 684–709.

Valentine, J. W. (1974). Temporal bias in extinctions among taxonomic categories. *J. Paleontol.*, **48**, 549–52.

Valentine, J. W. (1989). How good was the fossil record? Clues from the California Pleistocene. *Paleobiology*, **15**, 83–94.

Valentine, J. W. (1990). The macroevolution of clade shape. In *Causes of evolution*, (ed. R. M. Ross and W. D. Allmon), pp. 128–50. University of Chicago Press.

Valentine, J. W. and Jablonski, D. (1993). Fossil communities: Compositional variation at many time scales. In *Species diversity in ecological communities*, (ed. R. E. Ricklefs and D. Schluter), pp. 341–9. University of Chicago Press.

Van Valen, L. M. (1984). A resetting of Phanerozoic community evolution. *Nature*, **307**, 50–2.

Van Valen, L. M. (1985). How constant is extinction? *Evol. Theory*, **7**, 93–106.

Van Valkenburgh, B. and Janis, C. M. (1993). Historical diversity patterns in North American large herbivores and carnivores. In *Species diversity in ecological communities*, (ed. R. E. Ricklefs and D. Schluter), pp. 330–40. University of Chicago Press.

Ward, P. D. (1990). The Cretaceous/Tertiary extinctions in the marine realm; a 1990 perspective. *Geol. Soc. Am. Spec. Paper*, 247, 425–32.

Webb, T., III and Bartlein, P. J. (1992). Global changes during the last 3 million years: Climatic controls and biotic responses. *Ann. Rev. Ecol. Syst.*, **23**, 141–73.

Weems, R. E. (1992). The terminal Triassic catastrophic extinction event in perspective—a review of Carboniferous through early Jurassic terrestrial vertebrate extinction patterns. *Palaeogeog., Palaeoclimatol., Palaeoecol.*, **94**, 1–29.

Westrop, S. R. (1989). Macroevolutionary implications of mass extinction—evidence from an Upper Cambrian stage boundary. *Paleobiology*, **15**, 46–52.

Westrop, S. R. (1991). Intercontinental variation in mass extinction patterns: Influence of biogeographic structure. *Paleobiology*, **17**, 363–8.

Williams, G. C. (1992). *Natural selection*. Oxford University Press.

Winemiller, K. O. (1991). Ecomorphological diversification in lowland freshwater fish assemblages from five biotic regions. *Ecol. Monogr.*, **61**, 343–65.

Winston, J. E. (1992). Systematics and marine conservation. In *Systematics, ecology and the biodiversity crisis*, (ed. N. Eldredge), pp. 144–68. Columbia University Press, New York.

3

Constancy and change of life in the sea

J. B. C. Jackson

3.1 Introduction

Uniformatarianism is the null hypothesis of geology and gradual speciation by natural selection the null hypothesis of evolution. Ecology is the link between the two. The expectation is that gradual changes in earth processes drive gradual shifts in species and communities. Many ecological and evolutionary trends are in agreement with this scenario (Overpeck *et al.* 1992), but much new evidence suggests that changes in climate may be abrupt and unstable, with lasting effects on the subsequent climate and biota (Crowley and North 1988; Kennett and Stott 1991). Sudden environmental change may underlie the nearly universal occurrence of discontinuities in morphology and community composition which mark the age, stage, and epoch boundaries of the geologic timescale which traditionally have been attributed to the imperfections of the fossil record (Darwin 1859; Levinton 1988). Here I review evidence that communities of molluscs, reef corals, and planktonic Foraminifera have changed very little since the end of the Pliocene approximately 2 million years ago (mya). After at least 8 million years (my) of relative faunal stability, turnover of Late Pliocene faunas required only a few hundred thousand years (ky).

3.2 Patterns of speciation

There are now several palaeontological studies of patterns and rates of morphological change in the sea that allow distinction between punctuated and gradual speciation (Gould and Eldredge 1993). The best data are for Neogene to Recent marine invertebrates which have the advantage of an excellent fossil record and numerous fossil species that are still alive for comparison. Planktonic Foraminifera exhibit both gradual and punctuated speciation (Wei and Kennett 1988; Wei 1994) whereas cheilostome bryozoans are exclusively punctuated (Cheetham 1986; Jackson and Cheetham, in press). Most gastropod molluscs are punctuated, although there is one good example of gradualism (Michaux 1989; Geary 1990). Bivalve molluscs consistently exhibit morphological stasis but patterns and rates of changes in morphology at speciation have not been described (Stanley and Yang 1987). Based on these and other studies, cases of punctuation and stasis outnumber gradual evolution by more than ten to one for

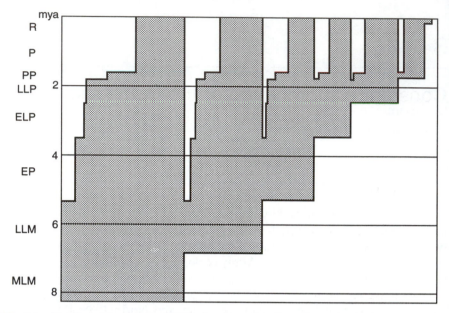

Fig. 3.1 Stratigraphic ranges of 395 subgenera of molluscs from the Caribbean coast of Costa Rica and western Panama. Timescale: R, Recent; 0, mya; P, Pleistocene; 0–1.6 mya; PP, Plio-Pleistocene boundary, 1.6–1.8 mya; LLP, late Late Pliocene, 1.8–2.5 mya; ELP, early Late Pliocene, 2.5–3.5 mya; EP, Early Pliocene, 3.5–5.3 mya; late Late Miocene, 5.3–6.8 mya (by interpolation); middle Late Miocene, 6.8–8.2 mya.

the over 100 species examined in detail. Thus, for all practical purposes, species of marine invertebrates exist as discrete packages in space and time whose durations and distributions can be usefully counted and compared.

3.3 The timing of speciation and extinction

Almost all living, well-skeletonized species of Foraminifera, corals, bivalves, and snails have persisted without obvious morphological change for 2 my and the majority for longer. This alone constitutes compelling evidence for evolutionary stasis in the sea, and demonstrates that speciation and extinction do not occur at random over time. The best data are for planktonic Foraminifera which have been sampled intensively world-wide as part of the Deep Sea Drilling Program (Kennett and Srinivasan 1983; Berggren *et al.* 1985). Median species durations for the 44 known species of Neogene globigeriniids is 10.4 my (range 1.0–34.9) and for the 50 globorotaliids is 4.4 my (range 1.0–23.7) (Stanley *et al.* 1988). Despite these long durations, however, 39% of globogerinid and 20% of globorotaliid species died out between 3 and 2 mya.

Molluscs from the Caribbean coast of western Panama and Costa Rica show a similar pattern (Figs 3.1 and 3.2; Jackson *et al.* 1993). The median duration of 395 genera and subgenera common enough to be analysed is 5.2 my (range 0.2–8.2 my

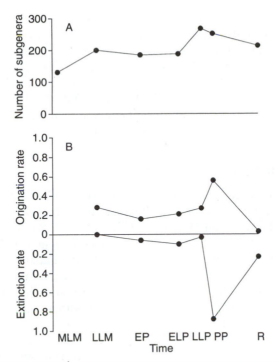

Fig. 3.2 Diversity, origination, and extinction of 395 subgenera of molluscs. Timescale as in Fig. 3.1. The origination rate at time t equals the number of taxa that first appear at t, divided by the total number of taxa at t, divided by the time between t and $t-1$. The extinction rate at t equals the number of taxa present at $t-1$ that are absent at t and thereafter, divided by the total taxa at $t-1$, divided by the time between t and $t-1$.

for all but 9 taxa known only from the Recent). Extinction rates were low until the end of the Pliocene about 2 mya when 18% of the taxa disappeared (Figs 3.1 and 3.2). Origination rate also doubled at the same time, so that altogether there was a 30% turnover of the fauna *at the generic and subgeneric level* within about 200 ky. There are no known younger (Pleistocene) fossil faunas in the region, so that the end Pliocene faunas can only be compared directly with the Recent. Nevertheless, 36% of these taxa went extinct some time during the Early Pleistocene, so that turnover lasted longer than 200 ky. Two-thirds of Early Pliocene species of molluscs from the southeastern United States also went extinct at the end of the Pliocene (Stanley and Campbell 1981; Stanley 1986), but were replaced by new species since diversity did not decline (Allmon *et al.* 1993). As for the Isthmus, the average durations of these younger species are unknown for lack of Pleistocene fossils. Such data are available for 747 fossil species from California, however, where the extinction rate over the last million years is only 13% (Valentine 1989; Valentine and Jablonski 1991).

Finally, Budd *et al.* (in press; submitted) revised the stratigraphy and taxonomy of 175 species of Caribbean reef corals over the past 24 my. Their median duration is approximately 7 my (range 150 ky–24 my). As for the other groups, there was a pulse of extinction at the end of the Pliocene, when extinction rates increased nearly 10-fold, and only 36% of the 106 species alive between 4–2 mya survived to the present. The only subsequent Pleistocene extinctions were of *Stylophora* and *Pocillopora*. Some other groups did not experience exceptional turnover or extinction at the end of the Pliocene. Only 6 of 107 common benthic Foraminifera from the Caribbean coast of Panama went extinct, and no common species evolved during the last 3 my (Collins submitted).

3.4 Evolution and environment

Many long-term evolutionary trends in the sea were probably due to interactions among species almost regardless of fluctuations in the physical environment (Vermeij 1987; Jackson and McKinney 1990). On the other hand, synchronous turnover of faunas is more closely tied to changes in climate and oceanographic conditions. The most likely explanation for the massive Plio-Pleistocene turnover of tropical western Atlantic Foraminifera, corals, and molluscs is climatic cooling due to the onset of northern hemisphere glaciation (Stanley and Campbell 1981; Stanley 1986; Stanley *et al.* 1988), with new estimates for tropical sea surface temperatures during glacial maxima of 5–6 °C below present values (Beck *et al.* 1992; Guilderson *et al.* 1994). Independent evidence for the cooling hypothesis comes from the geographic and habitat distributions of extinct and surviving species of bivalves and planktonic Foraminifera (Stanley 1986; Stanley *et al.* 1988) and altitudinal shifts in tropical terrestrial vegetation (Bush *et al.* 1992; Colinvaux submitted).

Differences in patterns and rates of Plio-Pleistocene turnover between oceans further implicate refrigeration as an important factor. Extinction of eastern Pacific molluscs at the end of the Pliocene was less than in the tropical western Atlantic (Jackson *et al.* 1993), and was preceded by different evolutionary trends in adult size and larval development (Jackson *et al.* submitted; Figs 3.3 and 3.4). Molluscan size and development vary with differences in primary production and food availability (Vance 1973; Vermeij 1978; Jablonski and Lutz 1983), and the data in Figs 3.3 and 3.4 agree well with inferred Pliocene increase in seasonality and upwelling in the eastern Pacific based on oxygen isotopic variation in mollusc shells, but more localized or ephemeral upwelling in the Caribbean (Teranes *et al.* submitted). Disappearance of well-developed eastern Pacific coral reefs after the Miocene (Budd 1989) also agrees with the inferred difference in seasonality between the two oceans.

Upwelling causes seasonal differences in temperature that may exceed 10 °C even in the tropics (D'Croz *et al.* 1991). Thus, Late Pliocene molluscs in the eastern Pacific were probably exposed to comparable seasonal temperature decreases that exceed estimates of a 5–6 °C drop in winter sea-surface temperatures due to glaciation. In contrast, Late Pliocene Caribbean species would not have been similarly pre-adapted to temperature decrease. Once past the initial thermal filter of glacial cooling, however, subsequent rapid rise and fall

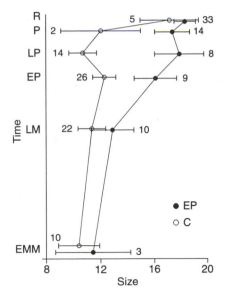

Fig. 3.3   Trends in strombinid shell size (square root height × width in mm) in the eastern Pacific (EP) and Caribbean (C). Numbers of species at each interval indicated alongside standard error bars. Ages: R, Recent; P, Pleistocene; LP, Late Pliocene; EP, Early Pliocene; LM, Late Miocene, EMM, Early and Middle Miocene combined.

of sea-level and temperatures had a negligible effect on Caribbean molluscan or reef coral extinction and diversity (Fig. 3.2; Budd *et al.* in press).

3.5   Prospects for marine diversity

Changes in marine species and communities are co-ordinated and discontinuous. This was as true 380 mya in the Devonian (Brett *et al.* 1990) as in the Neogene, and on the land (Vrba 1985) as in the sea. Species appear and disappear as morphologically discrete packages that come and go in surprising synchrony. Shifts in biotas are due to major changes in climate, but, once they occur, new biotas persist more or less intact for millions of years despite repeated occurrences of the same climatic fluctuations. Thus, continued Pleistocene cycles of low and high temperatures had little evolutionary effect (Potts 1984; Valentine 1989) because, once through the initial thermal filter, all that was left were eurythermal species.

What can these insights tell us about the prospects for marine diversity in the face of increasing human disturbance? Consider the case of Caribbean coral reefs where the abundance of live corals has decreased precipitously over the last 20 years with accompanying increase in macroalgae (Lessios 1988; Knowlton 1992). Caribbean reefs had been heavily exploited for at least 500 years previously with superficially little effect. Early good observers (Sloane 1701; Dampier 1968) described extraordinarily abundant populations of large vertebrates (manatees,

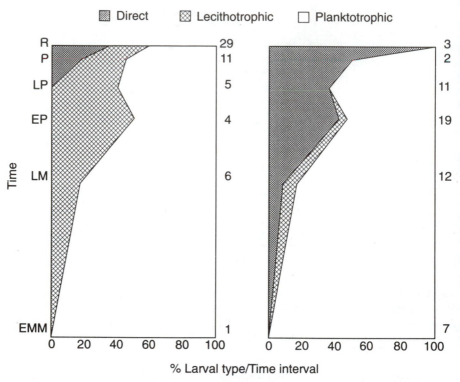

Fig. 3.4 Trends in relative frequency of species inferred to possess planktotrophic larvae (feed in plankton for several weeks), lecithotrophic larvae (non-feeding, drift in plankton about one week), and direct development (no free larval stage) in the eastern Pacific (left) and Caribbean (right). Numbers of species at each interval indicated to right of boxes for each ocean. Age intervals as in Fig. 3.3.

turtles, groupers) and molluscs (conchs, pearl oysters) that have not been seen anywhere in abundance for at least a century. Yet turtle was the *principal* supply of meat eaten in Jamaica during the 18th century (Catesby 1731; Long 1774) until they were virtually eliminated, just as the Maori eliminated the moas of New Zealand (Burney 1993). Manatees, turtles, jewfish, and conchs are not as dramatic as giant sloths and mammoths, but there is little doubt that they have been hunted to virtual extinction with the same efficiency.

As far as one can reconstruct from the Holocene record and the first detailed descriptions of reefs in the 1950s (Goreau 1959; Jackson 1992), the virtual disappearance of large vertebrates and conchs had little if any direct effect on the composition of Caribbean reef communities. Abundant smaller predatory and herbivorous fishes consumed sea-urchins, other invertebrate grazers, and algae, and the ratio of corals to macroalgae was still high (plane *a* in Fig. 3.5). Subsequently, intense subsistence fishing associated with rapidly rising human

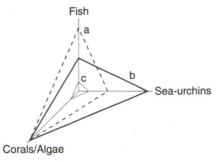

Fig. 3.5 Model of the response of Caribbean coral reef communities to over-fishing and sea-urchin disease. For details see text.

populations decreased both predatory and grazing fishes, resulting in a dramatic increase in invertebrate grazers such as the sea-urchin *Diadema antillarum* (Hay 1984). However, this sea-urchin also severely regulates algal growth, so that coral abundance remained high (plane *b* in Fig. 3.5).

But in 1983, mass mortality of *Diadema* (Lessios 1988) left the reefs without any efficient grazers. This sudden and dramatic shift in the relative abundance of different consumers and their prey permitted an enormous increase in unpalatable macroalgae that are progressively overgrowing corals throughout the Caribbean (plane *c* in Fig. 3.5), and reef communities appear to be at least temporarily locked into a macroalgal community state (Knowlton 1992). To make matters worse, increased nutrients due to the use of fertilizers, decreased coral abundance due to major storms, sedimentation associated with deforestation, toxins and oil spills, and mass bleaching of corals (Woodley *et al.* 1981; Cortes and Risk 1985; Tomascik and Sander 1987; Brown 1990; Hallock 1988; Guzman *et al.* 1991) all help to further lower thresholds for macroalgal dominance over corals (Knowlton 1992). At the current rate, there may be very few Caribbean corals left to be affected by a global climate change (Smith and Buddemeier 1992).

There is a disturbing parallel between this grisly story and the long-term persistence followed by rapid turnover described for Plio-Pleistocene marine communities. Threshold effects on the scale of entire biotas are irreversible.

Summary

Most marine species appear abruptly in the fossil record and persist unchanged for millions of years. Speciation and extinction commonly occur in pulses so that groups of species come and go as ecological units that dominate the seascape for millions of years. Dramatic turnover of mollusc, coral, and planktonic Foraminifera species occurred throughout tropical America about 2 million years ago in apparent response to the onset of northern hemisphere glaciation. In contrast, subsequent glacial cycles, temperature fluctuations, and sea-level change had little lasting biological effect. There is no necessary correlation between the magnitude of environmental change and the subsequent ecological and evolutionary response.

Acknowledgements

The ideas for this paper germinated during a sabbatical year at Wolfson College and the Department of Zoology, University of Oxford. Tony Coates, Tom Cronin, Helena Fortunato, Nancy Knowlton, and Paul Morris gave much helpful advice. Helena Fortunato and Xenia Guerra prepared the figures. This work was supported by the National Geographic Society, the Kugler Fund of the Basel Naturhistorisches Museum, National Science Foundation, and the Smithsonian Tropical Research Institute.

References

Allmon, W. D., Rosenberg, G., Portell, R. W., and Schindler, K. S. (1993). Diversity of coastal plain mollusks since the Pleistocene. *Science*, **260**, 1626–9.

Beck, J. W. *et al.* (1992). Sea-surface temperature from coral skeletal strontium/calcium ratios. *Science*, **257**, 644–7.

Berggren, W. A., Kent, D. V., Flynn, J. J., and Van Couvering, J. A. (1985). Cenozoic chronology. *Geol. Soc. Am. Bull.*, **96**, 1407–18.

Brett, C. E., Miller, K. B., and Baird, G. C. (1990). A temporal hierarchy of paleoecologic processes within a Middle Devonian epeiric sea. *Paleont. Soc. Spec. Publ.*, **5**, 178–209.

Brown, B. E. (ed.) (1990). Coral bleaching: special issue. *Coral Reefs*, **8**, 153–232.

Budd, A. F. (1989). Biogeography of Neogene Caribbean reef corals and its implications for the ancestry of eastern Pacific reef corals. *Mem. Ass. Australas, Palaeontol.*, **8**, 219–30.

Budd, A. F., Johnson, K. G., and Stemann, T. A. Stratigraphic distributions of genera and species of Neogene to Recent Caribbean reef corals. *J. Paleontol.*, **68**, in press.

Budd, A. F., Johnson, K. G., and Stemann, T. A. Plio-Pleistocene turnover and extinctions in the Caribbean reef coral fauna. In *Evolution and environment in tropical America*, (ed. J. B. C. Jackson, A. G. Coates, and A. F. Budd). University of Chicago Press, in press.

Burney, D. A. (1993). Recent animal extinctions: recipes for disaster. *Am. Sci.*, **81**, 530–41.

Bush, M. B. *et al.* (1992). A 14,300-yr paleoecological profile of a lowland tropical lake in Panama. *Ecol. Monogr.*, **62**, 251–75.

Catesby, N. (1731). *The natural history of Carolina, Florida and the Bahama islands*. Folio, London.

Cheetham, A. H. (1986). Tempo of evolution in a Neogene bryozoan: rates of morphologic change within and across species boundaries. *Paleobiology*, **12**, 190–202.

Colinvaux, P. A. Quaternary environmental history and forest diversity in the neotropics. In *Evolution and environment in tropical America*, (ed. J. B. C. Jackson, A. G. Coates, and A. F. Budd). University of Chicago Press, in press.

Collins, L. S. Environmental changes in Caribbean shallow waters relative to the closing tropical American seaway. In *Evolution and environment in tropical America*, (ed. J. B. C. Jackson, A. G. Coates, and A. F. Budd). University of Chicago Press, in press.

Cortes, J. and Risk, M. J. (1985). A reef under siltation stress: Cahuita, Costa Rica. *Bull. Mar. Sci.*, **36**, 339–56.

Crowley, T. J. and North, G. R. (1988). Abrupt climate change and extinction events in earth history. *Science*, **240**, 996–1002.

Dampier, W. (1968). *A new voyage around the world* (reprint). Dover, New York.

Darwin, C. (1859). *On the origin of species by natural selection*. John Murray, London.

D'Croz, L., Del Rosario, J. B., and Gomez, J. A. (1991). Upwelling and phytoplankton in the Bay of Panama. *Rev. Biol. Trop.*, **39**, 233–41.

Geary, D. H. (1990). Patterns of evolutionary tempo and mode in the radiation of *Melanopsis* (Gastropoda; Melanopsidae). *Paleobiology*, **16**, 492–511.

Goreau, T. F. (1959). The ecology of Jamaican coral reefs. I. Species composition and zonation. *Ecology*, **40**, 67–90.

Gould, S. J. and Eldredge, N. (1993). Punctuated equilibrium comes of age. *Nature*, **366**, 223–7.

Guilderson, T. P., Fairbanks, R. G., and Rubenstone, J. L. (1994). Tropical temperature variations since 20 000 years ago: modulating interhemispheric climate change. *Science*, **263**, 663–5.

Guzman, H. M., Jackson, J. B. C., and Weil, E. (1991). Short-term ecological consequences of a major oil spill on Panamanian subtidal reef corals. *Coral Reefs*, **10**, 1–12.

Hallock, P. (1988). The role of nutrient availability in bioerosion: consequences to carbonate buildups. *Palaeogeog., Palaeoclimatol., Palaeocol.*, **63**, 275–91.

Hay, M. E. (1984). Patterns of fish and urchin grazing on Caribbean coral reefs: are previous results typical? *Ecology*, **65**, 446–54.

Jablonski, D. and Lutz, R. A. (1983). Larval ecology of marine benthic invertebrates: paleobiological implications. *Biol. Rev.*, **58**, 21–89.

Jackson, J. B. C. (1992). Pleistocene perspectives on coral reef community structure. *Amer. Zool.*, **32**, 719–31.

Jackson, J. B. C. and Cheetham, A. H. (1994). Phylogeny reconstruction and the tempo of speciation in cheilostome Bryozoa. *Paleobiology*, **20**, in press.

Jackson, J. B. C. and McKinney, F. K. (1990). Ecological processes and progressive macroevolution of marine clonal benthos. In *Causes of evolution*, (ed. R. M. Ross and W. D. Allmon), pp. 173–209. University of Chicago Press.

Jackson, J. B. C., Jung, P., Coates, A. G., and Collins, L. S. (1993). Diversity and extinction of tropical American mollusks and emergence of the Isthmus of Panama. *Science*, **260**, 1624–6.

Jackson, J. B. C., Jung, P., and Fortunato, H. Paciphilia revisited: transisthmian evolution of the *Strombina* Group (Gastropoda: Columbellidae). In *Evolution and environment in tropical America*, (ed. J. B. C. Jackson, A. G. Coates, and A. F. Budd). University of Chicago Press, in press.

Kennett, J. P. and Srinivasan, M. S. (1983). *Neogene planktonic Foraminifera, a phylogenetic atlas*. Hutchinson Ross, Stroudsburg, Pennsylvania.

Kennett, J. P. and Stott, L. D. (1991). Abrupt deep-sea warming, paleoceanographic changes and benthic extinctions at the end of the Palaeocene. *Nature*, **353**, 225–9.

Knowlton, N. (1992). Thresholds and multiple stable states in coral reef community dynamics. *Amer. Zool.*, **32**, 674–82.

Lessios, H. A. (1988). Mass mortality of *Diadema antillarum* in the Caribbean: What have we learned? *Ann. Rev. Ecol. Syst.*, **19**, 371–93.

Levinton, J. (1988). *Genetics, paleontology, and macroevolution*. Cambridge University Press.

Long, E. (1774). *History of Jamaica, or general survey of the ancient and modern state of that island*. T. Loundes, London.

Michaux, B. (1989). Morphological variation of species through time. *Biol. J. Linn. Soc.*, **38**, 239–55.

Overpeck, J. T., Webb, R. S., and Webb, T., III (1992). Mapping eastern North American vegetation change of the past 18 ka: no-analogs and the future. *Geology*, **20**, 1071–4.

Potts, D. C. (1984). Generation times and the Quaternary evolution of reef-building corals. *Paleobiology*, **10**, 48–58.

Sloane, H. (1701). *A voyage to the islands Madera, Barbados, Neives, St. Christophers and Jamaica with the natural history of the herbs and trees, four-footed beasts, fishes, birds, insects, reptiles, etc. of the last of those islands*. Folio, London.

Smith, S. V. and Buddemeier, R. W. (1992). Global change and coral reef ecosystems. *Ann. Rev. Ecol. Syst.*, **23**, 89–118.

Stanley, S. M. (1986). Anatomy of a regional mass extinction: Plio-Pleistocene decimation of the western Atlantic bivalve fauna. *Palaios*, **1**, 17–36.

Stanley, S. M. and Campbell, L. A. (1981). Neogene mass extinction of western Atlantic molluscs. *Nature*, **293**, 457–9.

Stanley, S. M. and Yang, X. (1987). Approximate evolutionary stasis for bivalve morphology over millions of years: a multivariate, multilineage study. *Paleobiology*, **13**, 113–39.

Stanley, S. M., Wetmore, K. L., and Kennett, J. P. (1988). Macroevolutionary differences between the two major clades of Neogene planktonic foraminifera. *Paleobiology*, **14**, 235–49.

Teranes, J. L., Geary, D. H., and Bemis, B. E. The oxygen isotopic record of seasonality in Neogene bivalves from the Central American Isthus. In *Evolution and environment in tropical America*, (ed. J. B. C. Jackson, A. G. Coates, and A. F. Budd). University of Chicago Press, in press.

Tomascik, T. and Sander, F. (1987). Effects of eutrophication on reef-building corals. II. Structure of scleractinian coral communities on fringing reefs, Barbados, West Indies. *Mar. Biol.*, **94**, 53–75.

Valentine, J. W. (1989). How good was the fossil record? Clues from the California Pleistocene. *Paleobiology*, **15**, 83–94.

Valentine, J. W. and Jablonski, D. (1991). Biotic effects of sea level change: the Pleistocene test. *J. Geophys. Res.*, **96**, 6873–8.

Vance, R. R. (1973). On reproductive strategies in marine benthic invertebrates. *Amer. Nat.*, **107**, 339–52.

Vermeij, G. J. (1978). *Biogeography and adaptation*. Harvard University Press, Cambridge, MA.

Vermeij, G. J. (1987). *Evolution and escalation*. Princeton University Press, Princeton, NJ.

Vrba, E. S. (1985). African Bovidae: evolutionary events since the Miocene. *S. African J. Sci.*, **81**, 263–6.

Wei, K.-Y. (1994). Stratophenetic tracing of phylogeny using SIMCA pattern recognition technique: a case study of the late Neogene planktic Foraminifera *Globoconella* clade. *Paleobiology*, **20**, 52–65.

Wei, K.-Y. and Kennett, J. P. (1988). Phyletic gradualism and punctuated equilibrium in the late Neogene planktonic foraminiferal clade *Globoconella*. *Paleobiology*, **14**, 345–63.

Woodley, J. D. *et al.* (1981). Hurricane Allen's impact on Jamaican coral reefs. *Science*, **214**, 749–55.

4

Insect faunas in ice age environments: why so little extinction?

G. R. Coope

. . .and no one has the right to say that no water-babies exist till they have seen no water-babies existing; which is quite a different thing, mind, from not seeing water-babies.

(Charles Kingsley: *The water babies*)

4.1 Introduction

The above quotation highlights the problem that we face when trying to deal with extinction as a phenomenon, that is, the difficulty of dealing with negative evidence. How can we ever be certain that an organism has finally ceased to exist? It is easy enough at the individual level—death has its own particular finality about it—but at the racial level, recognizing extinction is a much more intractable problem. Everyone is justifiably convinced that the mammoths are globally extinct because such an enormous animal could hardly be overlooked. Yet, even here, the recently discovered dwarf mammoths that had survived on Wrangel Island in the Arctic Ocean until a mere 5000 years ago, show how inadequate the fossil record can be in dating the final demise of even such a conspicuous species as this. The rediscovery of the presumed extinct bird, the Takahe (*Notornis*) in New Zealand again shows how difficult it is to be certain that even large birds have finally become extinct. The recent finding of a totally new genus of antelope *Pseudoryx nghetinhensis* Mackinnon, in Vietnam shows how wrong it is to presume that all the large mammals on Earth are now 'known to science'. Extinction may be easy enough to define in theory but it is much more difficult to recognize in practice even among the most conspicuous inhabitants of our planet.

Among the very small and often cryptic animals, such as many species of insect, extinctions may be almost impossible to demonstrate in spite of our subjective conviction that a great many species are currently being exterminated by the burning of rainforests or drainage of marshlands. Even if we cannot quantify extinction rates under such circumstances, the fact that we are at this moment losing a wealth of biodiversity can hardly be questioned.

But how can we explore the ways in which insects respond to the challenge of drastic environmental changes? Perhaps one of the best ways to answer this question is by investigating the relatively recent geological past (the Quaternary period) when the frequent and intense climatic changes of the glacial/interglacial cycles provided just such an environmental challenge. The wealth of insect fossils (or rather subfossils) from this period yield abundant objective information on the way in which insects actually reacted to this challenge. In the light of the global-warming hypothesis we may ask, to what extent have our faunas (and floras) 'seen it all before'? Can they be expected to weather any coming storm in the same manner as they did in the past? Will the stratagems that they adopted then in response to natural hazards, suitably equip them to survive those of human origin? Does the study of the recent past provide us with a key to understanding the future survival, or otherwise, of our present-day flora and fauna in the face of expected environmental changes?

4.2 Quaternary environmental change

In order to understand the structure and development of the Earth's biota of the present day it has to be set against the backdrop of the extraordinary events of Quaternary environmental changes. During the last 2 million years there have been numerous and intense climatic oscillations that were widespread and as severe, or even more so, than any recorded in the earlier parts of geological history. Because of the proximity of this period to the present day, it has been possible to document these changes in more detail, to quantify the changes involved and to provide a more precise chronology of these events than can be matched in the more remote parts of the stratigraphical history.

For over a century the sequence of Quaternary glaciations and their intervening interglacials has had to be interpreted from the patchy remains of deposits laid down on land, in an environment that was dominated by erosional processes. The record was thus incomplete, with major lacunae throughout and with geographically widespread remnants that were difficult to correlate with one another. The general tendency, therefore, was to underestimate the number of climatic events. Classically, there were said to have been four major cycles of glacials and interglacials with occasional minor warm episodes within the cold periods that were termed interstadials.

This overall simplistic view has changed completely with the advances in technology that enabled sedimentary cores to be retrieved from the ocean depths. The sediments that accumulated on the ocean floor might be expected to preserve a continuous record of events unaffected by the problems that afflicted the terrestrial sequences. Pioneer work by Shackleton (1977) showed that the variations in the ratio of ^{18}O to ^{16}O in benthonic foraminiferal tests at different depths in the sediment core provided an indication of the changes in the isotopic composition of the ocean water which in turn was a reflection of changes in global ice volume. Since the cores could be shown to extend back in time as far as and further than the last major reversal in the Earths magnetic field, dated at 730 000 years ago, it was possible to place the fluctuations in global ice volume into a secure time domain. Herein lay the makings of a true scientific revolution. The numbers of recognized climatic events greatly increased (see Fig. 4.1) and it became evident that for the last three-quarters of a million years at least, the Earth's normal climatic mode has been one of glacial conditions rather than the interglacial regime that prevails at the present day, and which we tacitly assume is the usual state of affairs. Thus, the climate over the past half million years has been as good as, or better than, the present climate for only about 2% of the time. Furthermore, Fig. 4.1 shows that many of the transitions between glacial and interglacial climatic modes were sudden and not gradual as had been hitherto believed. This was apparent even though the mixing time of the ocean waters is about 1000 years, meaning that the marine isotopic record can not resolve changes that took place in shorter periods than this. In fact, climatic changes could have been even faster than those shown in Fig. 4.1.

A similar picture of frequent and rapid changes has been obtained from studies of planktonic foraminifera from a transect of cores taken in the North Atlantic Ocean (Ruddiman and McIntyre 1981). They show that there were sudden changes in the patterns of circulation of ocean surface water and that the present-day position of the North Atlantic Drift (the Gulf Stream) that carries warm water northwards up the west coast of the British Isles and which is the moderator of our climate today, is only a relatively short-term interglacial phenomenon. During the more prolonged glacial periods this current flowed almost from east to west at about lat 45 °N and turned southward at the latitude of central Portugal. At such times the circulation of the water in the northern Atlantic Ocean was dominated by a large clockwise gyre that brought polar water with its attendant floating pack-ice to our western seaboard. The climatic implications of these circulatory changes were intense for western Europe but they also reflected more widespread global changes. The dramatic switch from a glacial to interglacial circulatory pattern would seem to have taken place in an irresolvably short period of time (Ruddiman and McIntyre 1981).

Rapid switches in climatic signals are also recorded from isotopic changes in the cores drilled to a depth of more than 2 kilometres into the Greenland ice-cap. Here, however, it has been possible to obtain a much more precise measure of the times involved and thus the rates of climatic change, by counting the annual ice-layers. Dansgaard et al. (1989) showed that the transition from glacial to interglacial climate at the close of the Last Glaciation took place in about 50 years.

At least some of the frequent, sudden, and intense climatic oscillations recorded in the Greenland ice-cores extended into lower latitudes and have been known for many decades in western Europe where they have been recognized from changes in the pollen flora and insect faunas preserved in detritus peats that accumulated in lakes and pools. This flora and fauna evidence shows, for instance, that the climate was very unstable during the transition from the last glacial to the present-day interglacial period. To give an example, the sequence of climatic events at this time in southern Britain may be summarized as follows. It must be born in mind, however, that the timing adopted here is measured in

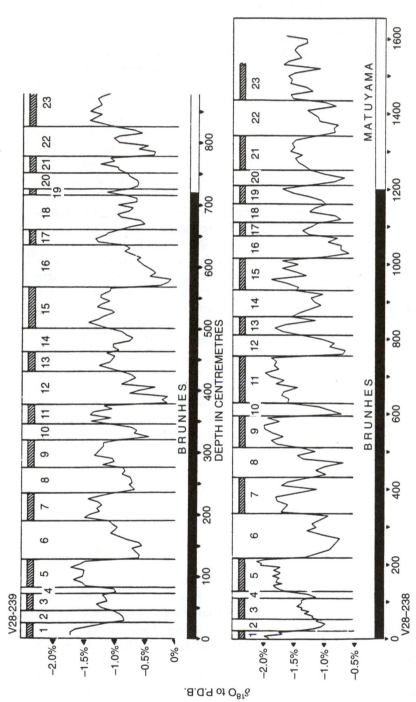

Fig. 4.1 Oxygen isotope records for cores V28–238 (bottom) and V28–239 (top) based on benthonic foraminifera. These levels are largely controlled by variation in global ice volume. The numbers along the top of the curves are the isotopic stages in which odd numbers have been allocated to periods of relatively low ice volume (i.e., more temperate periods) and even numbers allocated to period of relatively high ice volume (i.e., more glacial periods). The Brunhes–Matuyama magnetic reversal boundary in dated at c. 735 000 years ago and is situated in isotope stage 19. The present-day interglacial is represented by isotope stage 1. (From Shackleton 1977.)

radiocarbon years before present and not in calendar years. At about 13 thousand years ago the climate suddenly ameliorated so that mean July temperatures rose from about 10 °C to about 17 °C in less than a century. A rate of change of at least 1 °C per decade has been suggested for this temperature change (Coope and Brophy 1972). At this time, the change from fully glacial to fully interglacial climatic conditions was completed within the span of one human lifetime. Almost immediately after this temperate interlude, which probably lasted for only a few centuries, the summer warmth began to decline again with winter temperatures cooling even more rapidly. It now seems probable that this decline continued in rather a sporadic fashion, with several minor oscillations on the way, until at about 11 thousand years ago a rather more severe drop in temperature once more ushered in a return of glacial conditions with mean July temperatures at, or even below, 10 °C and the mean temperature of the coldest month of the winter at about − 15 °C. There was then a second sudden and intense climatic warming, this time dating from about 10 thousand years ago, when summer temperatures again rose very rapidly so that the climate returned to full interglacial warmth in a matter of a century or so (Ashworth 1972, 1973; Osborne 1973). Winter temperatures at this time again rose at a faster rate than those of the summer as the climate changed from a cold continental regime to the temperate oceanic conditions of the present day. This summary gives some idea of the major climatic challenges that the ancestors of the present-day biota had to face in their relatively recent geological past. Numerous synopses deal with the more detailed response of the various components of the biota to the Quaternary environmental changes (e.g., Lowe and Walker 1977; Bowen 1978).

At first sight it might be expected that widespread extinctions would result from such frequent, intense, and rapid climatic changes, with their attendant expansions and contractions of great ice-sheets covering large continental areas and consequent global sea-level variations of more than 100 metres. The mass extermination of so many of the large mammals and birds during the Quaternary period would appear to support this expectation. Yet, the less well-known and certainly less spectacular insect fossils tell a very different story.

4.3 Quaternary fossil insects

The only objective evidence as to which animals and plants existed in the prehistoric past and where they lived, is fossil evidence. On the basis of the biology of living organisms alone, it is possible to make inferences about biogeographical history, evolutionary change or even extinction rates, but all such hypotheses must remain conjectural. In the end, it is the test of palaeontology that determines how close these interpretations are to reality. Fortunately, the insects have left abundant fossil evidence from most phases of the great Quaternary climatic episodes.

The insect fossils that we are concerned with here may be found in great abundance in a wide variety of sediments, provided that these deposits have remained more or less waterlogged and preferably anoxic since the time of their deposition (oxidation rapidly destroys all insect fossils). Suitable localities include lake basins or even relatively small pools into which terrestrial debris has been continuously washed from the adjacent land surface. Mires that form in gradually subsiding depressions, such as solution hollows or within tectonic grabens, accumulate detritus muds that can often provide long histories of the swamp communities. Places such as these may preserve fairly continuous sequences that date back over hundreds of thousands of years. Overbank deposits on the flood plains of ancient rivers, left behind at the present day as terrace remnants, can also extend the record back in time hundreds of thousands of years, although in such cases, the sequences are discontinuous and represent short-term events only. Many peat bogs provide a continuous inventory of the fauna and flora over the last 10 thousand years and archaeological excavations in wet deposits can bring the story up to the threshold of historic time and the commencement of scientific record-keeping.

Unlike most of the fossils of classical palaeontology, Quaternary insect fossils are not mineralized but are represented by the original chitin of their exoskeletons. Because present-day insect taxonomy is largely based on exoskeletal characters and because we are dealing with the original material, it is possible to make precise comparisons of the fossils with modern specimens. Thus, the fossils can be identified using exactly the same criteria that are employed by entomologists to differentiate between present-day species; exactly the same species concept can be used in both entomology and palaeoentomology and the same nomenclature adopted for both modern and Quaternary specimens.

In the discussion that follows, most of the conclusions will be drawn from the Coleoptera (beetles) simply because their robust skeletons make such good fossils. It must be emphasized, however, that beetles are unlikely to be exceptional with regard to their evolutionary or biogeographical histories and that they can stand as proxies for other orders of insects not so well blessed with such an abundant fossil record. Some of these orders are well represented in our fossil assemblages but are rather more difficult to study or have been overshadowed by the more conspicuous and exciting Coleoptera. Thus the Diptera are particularly abundant but often in a very crushed state of preservation but chironomid larval heads are often sufficiently common that they can be analysed statistically to give significant evidence on past limnic environments (Walker 1987). Hymenoptera are represented by numerous highly distinctive skeletal elements that still await serious investigation. Hemiptera are fairly common in some deposits and would certainly merit specialist attention. Trichoptera are probably the earliest colonizers of lakes and ponds after the retreat of ice-sheets and their larval sclerites provide much palaeolimnological data (Williams 1988; Wilkinson 1984). Odonata are represented only by isolated segments and the central part of the heads of damsel flies. Orthoptera are completely absent from our fossil assemblages, as are the Lepidoptera, probably because of the frailty of their exoskeletons. Caterpillar jaws are the only evidence that we have of the original presence of the latter.

The literature on Quaternary entomology is widely scattered and often rather inaccessible in the literature of many branches of science. We have, therefore, produced a bibliography and literature review (Buckland and Coope 1991). In this review it will be seen that most of the early workers on Quaternary insects, up to about the 1950s, attributed them to new species, basing their interpretations on the notion that speciation was widespread and common among insects at the present day and that the fossils, dating from tens or even hundreds of thousands of years ago, must therefore represent extinct species that had either died out or evolved into something else. It was the challenge of the ever-increasing complexity of the Quaternary climatic changes that provided ample justification for this expectation and the immense diversity of insect species reinforced this sense of rapid, ongoing evolution.

This traditional view that successive glacials and interglacials caused widespread extinctions of insect species and complementary speciation cannot now be sustained. In the last few decades, intensive investigations of Quaternary insect fossils in Britain, Sweden, Denmark, France, Russia, and in North and South America have all shown that species of Coleoptera remained constant both in their morphology and in their environmental requirements throughout the whole of the Quaternary period (Coope 1970). From the British Isles alone there are now over two thousand fossil species known that are the precise match of their present-day equivalents. What is more, there is no large backlog of distinctive but unidentified fossil material that could represent the remains of extinct forms. Certainly, the number of mystery species that defy identification and may possibly be extinct species is very much less than 1% of the total.

Highlighted below are a selection of species that have been discovered as fossils in Quaternary deposits in Britain but which, when first discovered, did not resemble any familiar species. Originally, these species seemed to be suitable candidates to be viewed as extinct species. However, each has subsequently been found living at the present day, albeit in rather unexpected parts of the world.

1. *Carabus maeander* Fisch. This is a North American and Asiatic species whose present-day range in Siberia reaches as far west as the Lena River. It has been found as a fossil in Britain at two localities in deposits that date from the Upton Warren Interstadial in the middle of the Last Glaciation (Devensian, Weichselian, Wurm) (Coope 1962; Briggs *et al.* 1985).

2. *Helophorus obscurellus* Popp. This species is a predominantly Siberian species that lives today on the tundras that border the Arctic Ocean and also on the high cold steppes of central Asia. Its nearest locality to Britain is on the Kanin peninsula in arctic Russia. As a fossil this species has been found commonly in Britain and on the adjacent continent, in deposits that were laid down under cold conditions. It only became extinct in Britain about 10 thousand years ago, at the time of the sudden climatic warming at the start of the Postglacial period. The fossils were originally described as *Helophorus wandereri* by d'Orchymont in 1927, in the belief that they represented an extinct species, but were recognized as being representative of a living species by R. B. Angus (in Coope 1968).

3. *Helophorus arcticus* Brown. This is an almost exclusively arctic North American species that has recently been recorded from the north west of the Kamchatka peninsula in eastern Siberia (R. B. Angus, pers. comm., 1992). As a fossil it has been recovered from Middle Pleistocene deposits at Ardleigh, Essex (Coope, unpublished data).

4. *Micropeplus dokuchaevi* Rjaburkhin. This species has a significant taxonomic history. It was first described by J. V. Matthews, Jr. (1970) from Pliocene deposits in Alaska that were sealed by a lava flow from which a date of 5.7 million years was obtained. He named the fossils *Micropeplus hoogendorni* believing that they represented an extinct species that could have been the precursor of some present-day member of this genus. Matthews' species has subsequently been equated with the living Siberian species *Micropeplus dokuchaevi* Rja. (opinion of M. J. Campbell in Elias 1994, p. 59). As a fossil it has been found in three Middle Pleistocene sites in the English Midlands (Shotton *et al.* 1993).

5. *Oxytelus* (*Anotylus*) *gibbulus* Epp. At the present day this species is almost entirely confined to the Caucasus Mountains, but there is a single enigmatic record from the Ussuri region of eastern Siberia. (Hammond *et al.* 1979). As a fossil this species was at times the most abundant staphylinid beetle in deposits from a number of sites in the English Midlands dating from the as yet unnamed interglacial that appears to be equivalent in time to 'isotope stage 7' (see Fig. 4.1), namely from about 200 thousand years ago. It continued to occur sporadically in Britain up to the Upton Warren Interstadial in the middle of the Last Glaciation (Coope *et al.* 1961) where the distinctive male head was illustrated in the hope that, if it was still a living species, someone might recognize it. It was identified by W. O. Steel (in Coope 1970).

6. *Tachinus jacuticus* Popp. Today, this species is almost totally confined to Siberia with only a single outpost in north-eastern Russia. It occurs in Britain as a common fossil in deposits that accumulated during the middle phases of the Last Glaciation (Coope *et al.* 1961). It recolonized the British Isles as the ice-sheets finally retreated, only becoming extinct here about 10 thousand years ago when the climate suddenly ameliorated at the beginning of the present interglacial. In this case also the fossils were illustrated in the hope that they would be recognized and they were identified by Ullrich and Coope (1974).

7. *Tachinus caelatus* Ullrich. At the present day this species appears to be restricted to the mountains near Ulan Bator in Mongolia (Ullrich 1975). As a fossil it has been found frequently in England in deposits that were laid down under cold continental climatic conditions during various periods of the Middle Pleistocene (Taylor and Coope 1985).

8. *Aphodius holdereri* Reitt. Today this species is confined to the high plateau of Tibet and the adjacent parts of north-western China. As a fossil this species was

the most abundant dung beetle in Britain during the colder parts of the Last Glaciation (Coope 1973).

These examples serve as cautionary tales. Many others could be drawn from the British Quaternary fossil record, but they would all have carried the same message, namely, that there is no evidence to support the notion that widespread extinctions of insect species resulted from the numerous intense and rapid climatic changes of the last 2 million years. These changes certainly caused numerous local, often catastrophic extinction events, but, curiously enough, global extinctions were apparently very rare.

To illustrate further the ways in which geographical distributions of insect species have altered in both space and time, even within the relatively short period of the latest glacial/interglacial cycle, six species have been selected to represent the shifts in range of 'northern' and 'southern' species. Figures 4.2–4.7 show the locations of fossil occurrences in the British Isles of each species together with a small inset map of its present-day European distribution. Three species have been chosen from cold climate sites and three from temperate ones.

Figures 4.2, 4.3, and 4.4 are of northern species that lived in Britain well to the south of their present range during cold climate episodes of the Last Glaciation. *Diacheila polita* is a flightless species that was common in Britain during the middle period of the Last Glaciation. It and other similarly flightless species, apparently failed to recolonize Britain during the time of unstable climates at the close of this glacial period when the rapid climatic changes demanded great geographical agility. In contrast, *Diacheila arctica* and *Boreaphilus henningianus* are both capable of flight and were thus able to respond more quickly to any sudden climatic changes. Thus, both were able to return to Britain during the period of climatic instability during the Lateglacial cold phase (i.e., between 12 000 and 10 000 years ago), when mobility would be at a premium.

Figures 4.5, 4.6, and 4.7 are southern species that were able to extend their ranges northwards into Britain during the short temperate phase of the Lateglacial (i.e., between 13 000 and 12 000 years ago). Of these, *Bembidion octomaculatum* only just qualifies as a British species, seemingly unable to establish permanent breeding colonies even in the extreme south of England, though it is evidently capable of flying in from the continent. As a Lateglacial fossil it occurs in such numbers in the north of England that it was almost certainly breeding here then. *Bembidion grisvardi/ibericum* and *Asaphidion cyanicorne* are both capable of flight and must have responded almost instantly to the sudden climatic warming.

It should be emphasized, however, that although some of these 'northern' and 'southern' species occurred at the same sites (see Figs 4.2–4.7), they did not coexist in the British Isles. The apparent coincidence of some of the find spots is because most of these localities include sequences of sediments spanning several thousand years.

These figures also indicate the accumulated wealth of data on the biogeographical history of insect species that is gradually being built up from Quaternary

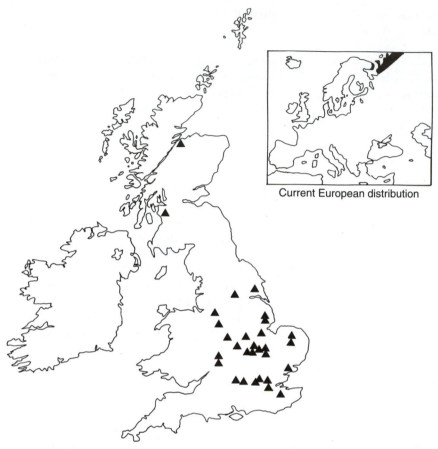

Current European distribution

Fig. 4.2 Fossil sites in the British Isles for *Diacheila polita* Fald. ▲, Devensian (Weichselian) Glacial and Interstadial sites. There are no Lateglacial records from the British Isles.

fossils. It would be impossible to infer much of this history by reference to their present-day distributions alone.

4.4 Discussion

We are now in a position to answer the first of the questions posed earlier in this chapter; namely, how did insects in the past respond to the challenge of drastic environmental change? Species have three possible responses to the large-scale climatic changes: (1) they may evolve out of trouble; (2) they may move out of trouble; or (3) they may become extinct. It was pointed out earlier in this chapter that the rates of change in climates have, in the past, been dramatically sudden. The rapidity of such changes makes adaptation by genetic change almost impossible. However, the mobility of insects is such that they can readily alter their ranges as the climate changes. The location of suitable climatic areas can

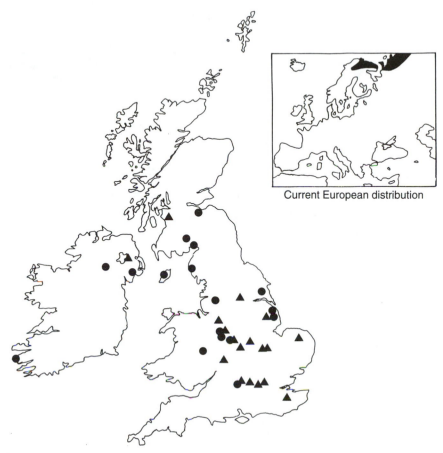

Current European distribution

Fig. 4.3 Fossil sites in the British Isles for *Diacheila arctica* Fald. ▲, Devensian (Weichselian) Glacial and Interstadial sites; ●, Devensian (Weichselian) Lateglacial cold sites.

thus be tracked from place to place across the continents. Paradoxically, even though the climate changed rapidly and intensely, the actual conditions in which a species lives remained effectively the same. Looked at from the species point of view, it was the geography that changed. Thus, providing that the mobility of a species is adequate and space is available, there should be no greater risk of extinction just because the climate swings back and forth between glacial and interglacial extremes.

If, however, the mobility of a species were reduced for some reason, say because its photo-period requirements made it impossible to change latitudes, the necessary tracking of acceptable climates might become impossible and extinction a likely outcome. To some extent this hypothesis can be tested. Since the major climatic oscillations in the northern hemisphere appear to have begun very suddenly at 2.47 million years ago (Ruddiman and Raymo 1988), the

Fig. 4.4 Fossil sites in the British Isles for *Boreaphilus henningianus* Sahlb. ▲, Devensian (Weichselian) Glacial and Interstadial sites; ●, Devensian (Weichselian) Lateglacial cold sites.

obligation for thermally sensitive species to track suitable climates must have been initiated at about the same time. It is therefore interesting to observe that in the studies of the Upper Tertiary insect faunas from Alaska and arctic Canada by John Matthews (1974, 1976) and from the far north of Greenland by Jens Böcher (1986), amongst many fossil species that are identical to modern ones, there are a few that would seem to be genuinely extinct and furthermore they do not seem to have survived into the Quaternary period. These late Tertiary extinctions may represent species that failed to track the appropriate conditions when the Quaternary climatic oscillations begun. A corollary to this hypothesis might be that any species that could survive the first major climatic challenge, could, by adopting the same stratagem, survive all the succeeding ones. It is as if we viewed Figure 4.1 as a hurdle race in which any species that could leap the first gate could similarly leap the rest.

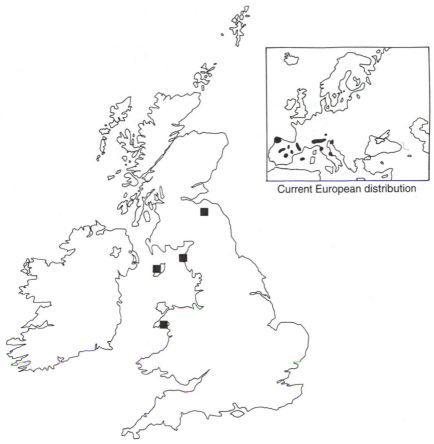

Current European distribution

Fig. 4.5 Fossil sites in the British Isles for *Asaphidion cyanicorne* Pand. ■, Devensian (Weichselian) Lateglacial temperate site.

Another way in which species mobility could be reduced is by isolation on an island from which escape is impossible. An oceanic island or equatorial mountain top would suffice. Species trapped on mountains in middle latitudes may have greater mobility than seems likely at first sight because it is always possible for them to be released from their confinement from time to time during the glacial periods. Under conditions of long-term isolation a species must endure any change on the spot. If the option of movement is denied it, only two alternative responses remain to a species: to evolve out of trouble or to become extinct. Both these responses would seem to be unusual in the light of what is now known about the way the bulk of insect species on continental masses reacted to major climatic changes. The maintenance of long-term species constancy and low extinction rate may thus be a 'mainland effect' in contrast to rapid evolution and high extinction rates that characterize islands. It is important not to extrapolate from the one to the other.

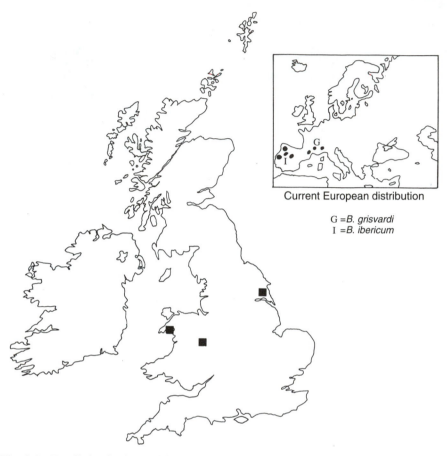

Current European distribution

G = *B. grisvardi*
I = *B. ibericum*

Fig. 4.6 Fossil sites in the British Isles for *Bembidion grisvardi* Dew. or *ibericum* Pioc. ■, Devensian (Weichselian) Lateglacial temperate site.

The apparent paradox of the prolonged constancy of insect species versus the extreme inconstancy of the Quaternary climate may be resolved as follows (Coope 1978). The high frequency, intensity, and suddenness of these climatic oscillations, coupled with the great mobility of insect populations, means that the geographical ranges of species are essentially transitory stages in a state of constant dynamic flux as they responded to the ever shifting climates. The large-scale tracking of acceptable environments means that populations continuously split up and reform as they progress to and fro across a complex landscape. The numerous episodes of temporary isolation and subsequent coalescence ensure that the gene pools were kept well stirred. Sustained evolution under such circumstances must have been almost impossible. Thus, paradoxically, it is the climatic inconstancy that can be seen to be an important factor, in the maintenance of specific constancy.

The ability to track acceptable climates from place to place across the

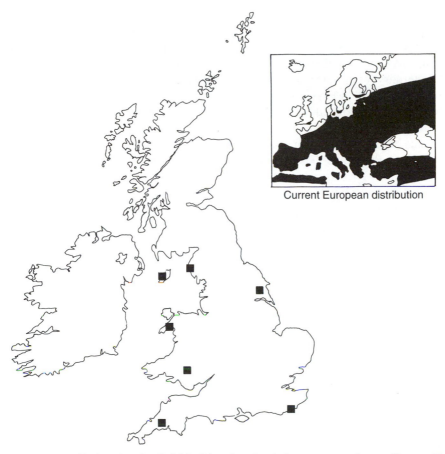

Fig. 4.7 Fossil sites in the British Isles for *Bembidion octomaculatum* Goeze. ■, Devensian (Weichselian) Lateglacial temperate site.

continental masses has another evolutionary consequence. If the environment in which a species lived may remained effectively constant regardless of the climatic changes, species constancy could have been reinforced by stabilizing selection to fit that particular environment for millions of generations; throughout the whole Quaternary period at least, regardless of its highly unstable climates.

There is a problem here of credibility. How can stabilizing selection be expected to operate with such precision as to maintain specific constancy in all its intimate detail for the several millions of generations represented by the Quaternary fossil record? This is a problem that confronts all evolutionary biologists when faced with the longevity of so many fossil species. In the case of the fossil insects discussed here, this problem can not be avoided or explained away on the basis that palaeontologists and neontologists have such different concepts of species that we are not comparing like with like. The Quaternary fossil insects show constancy in exactly the same features that are used to differentiate the present-day species. The problem of the mechanism whereby species stability is maintained, remains to be resolved. *The fact* of specific constancy is beyond question in all those regions of the world from which Quaternary fossils have been obtained. However, there are notable gaps in our record. It would be intriguing to know if a similar degree of species constancy was to be found in more equatorial regions, but as yet, there is no Quaternary fossil insect evidence from these latitudes. There are no fossil insects from tropical islands but surely conditions must have been suitable for their preservation there too. Here, there must be opportunities to test all sorts of treasured hypotheses.

If we are to guess (predict is too strong a word) the chances of survival of insect species in any climatic changes in the near future—in conditions of global warming for instance—more information is required on the way, or ways, in which species stability was maintained for several millions of generations in the geologically immediate past. Some speculations on the mechanism governing this long-term species constancy are therefore inescapable.

It is well known that insects are neurologically 'hard-wired' compared, for instance, with most vertebrates, a condition that also seems manifest in their biochemistry and behaviour patterns. Rigid adaptations would make any deviation from the *status quo* highly precarious and speciation a profound risk. It is also well known that natural populations are exceedingly heterozygous, suggesting that the stereotyped wild individual is sustained by a genetic complex in such a way that a wide variety of different genetic combinations result in the emergence of very similar 'wild types' (an old-fashioned term but none the less valuable). This wild phenotype would seem to have been honed for many millions of generations to fit a narrowly defined ecological niche. Here, we appear to have another paradox in which sameness of phenotype would seem to be maintained by an elaborate conspiracy of polygenic controls. In situations where there is reduced heterozygosity, the genetic constraints that maintain species constancy would become weakened, increasing the likelihood of novelties being expressed in the phenotype and thus made accessible to natural selection. Phenotypic diversity can thus be the result of moderate impoverishment of the gene pool (early stages of domestication probably fall into this category). The ultimate in impoverishment (i.e., homozygosity) will, of course, result in a return to phenotypic monotony (late stages of domestication fall into the latter category).

We are now in a position to contribute some tentative solutions to the second question posed at the start of this chapter, namely, will the stratagems so successfully adopted by insect species in the past to deal with glacial/interglacial climatic changes, suitably equip them to survive any impending humanly induced climatic changes? Two important factors have to be borne in mind. First, there is the demonstrable ability of insect species to track the climatic changes, thus freeing them from the necessity to adapt afresh to each successive climatic oscillation. Second, there is the proven ability of insect species to maintain constancy over millions of generations. Both these factors, although fitting the species well enough for survival in the climatic hurly-burly of the geologically recent past, may be of little value in the humanly contrived environmental chaos of the near future.

At the present time, human impact on the landscape is on an unprecedented scale. On the negative side there is the drainage of marshlands or the destruction of ancient mature forests, and on the positive side, there is the widespread blanket of intensive agriculture or the corduroy plantations of monotypic forestry, all of which make tracking of acceptable environments difficult or impossible for all but a few pest species. The effects of habitat destruction on insect species have been considered by Thomas and Morris (Chapter 8) listing losses in recent years and by Buckland *et al.* (1993) drawing on fossil evidence of post-glacial extinctions in Britain. The lists are impressive and depressing.

Bearing these recent extinctions in mind, how might these losses be minimized in the near future? The establishment of nature reserves provide the only refuges for many vulnerable species. However, these are currently conceived as essentially static in their locations, designed to protect habitats and species wherever they happen to be at the moment. No provision is made for any anticipated climatic change. They are prisons, guarded by wardens, in a sea of environmental hostility; as isolated as if they were true ocean islands. Since, for all flightless species, escape is impossible, they must now endure any change on the spot. With their erstwhile stratagem of moving out of trouble denied them and long-term stability so built into their make-up, extinction seems inevitable. We are creating a novel situation by the establishment of isolated islands in the middle of continents in which their faunas, and floras, may prove to be just as much at risk as those of remote Pacific islands.

It would be out of place here to discuss a methodology for the redesigning of nature reserves but a few observations seem necessary if widespread extinctions of our insect fauna are not to continue. Reserves should be large enough to maintain a complex of habitats, with populations of adequate size to ensure species stability. Even though some gene pools will inevitably be greatly reduced, it seems highly unlikely that this reduction in size and the consequent relaxation of the constraints that maintain species constancy will result in adequate evolutionary opportunities to make up for the losses by extinction. To enable species to track the changes in acceptable climates as they did in the recent past, reserves should be made as part of an extensive network connected, by corridors along which species can move freely from one reserve to another as changing climates dictate. It is, of course, difficult to envisage how such a network could be achieved in a country as overdeveloped as Britain, unless new motorways (or even railways) were designed to involve integral planning for linear nature reserves alongside, utilizing blighted properties, isolated farm fields or wood-lands, or even marshlands impounded by the motorway itself. These new reserves would have to be purpose built to maintain biotal diversity, complexity and continuity. Without plans such as these, designed to take into account the ways in which insect species have reacted to environmental challenges in the past, it seems inevitable that their rates of extinction will become catastrophic on a truly unprecedented scale.

Summary

The extinction of species of small invertebrates is difficult to recognize. However, in deposits that date from the last few million years, insect fossils are remarkably common and provide objective data on the history of the organisms that constitute the biotic communities of the present day. It might have been expected that the great climatic oscillations of the glacial/interglacial cycles should have caused widespread extinctions, if their effects on the large vertebrates are taken as our model. Yet the record of Quaternary fossil insects does not provide any evidence for high rates of extinction during this period. Furthermore, there is no evidence of widespread evolutionary change during this period and long-term constancy of species and communities of species, can be demonstrated to be the norm for at least the last million or so years (for many species one year is equivalent to one generation).

The enigma of how such constancy was sustained in the face of large-scale climatic fluctuations is an intriguing puzzle and several possible solutions are suggested as to why it should be so. These solutions themselves carry implications of significance in our estimates of present-day and future extinction rates of insect species in the light of humanly induced environmental changes. The stratagems that they employed so successfully in surviving the numerous, intense, and sudden climatic changes of the past may be of little value to them in their efforts to survive the human-induced changes of the near future.

References

Ashworth, A. C. (1972). A late-glacial insect fauna from Red Moss, Lancashire, England. *Entomologica Scandinavica*, **3**, 211–24.

Ashworth, A. C. (1973). The climatic significance of a late Quaternary insect fauna from Rodbaston Hall, Staffordshire, England. *Entomologica Scandinavica*, **4**, 191–205.

Böcher, J. (1986). Boreal insects in northernmost Greenland: palaeoentomological evidence from the Kap Kobemhavn Formation (Plio-Pleistocene) Peary Land. *Fauna Norvegica*, **B36**, 37–43.

Bowen, D. Q. (1978). *Quaternary geology: a stratigraphic framework for multidisciplinary work*, pp. 1–221. Pergamon, Oxford.

Briggs, D. J., Coope, G. R., and Gilbertson, D. D. (1985). *The chronology and environmental framework for early Man in the Upper Thames Valley.* British Archaeological Reports, Vol. 137, pp. 1–176, Oxford.

Buckland, P. C. and Coope, G. R. (1991). *A bibliography and literature review of Quaternary entomology*, pp. 1–85. J. R. Collis Publications, University of Sheffield.

Buckland, P. C. and Dinnin, M. H. (1993). Holocene woodlands, fossil insect evidence. In *Deadwood matters: the ecology and conservation of saproxylic environments in Britain*, (ed. K. J. Kirby and C. M. Drake), pp. 6–20. English Nature Science Publications No. 7, Peterborough.

Coope, G. R. (1962). A Pleistocene coleopterous fauna with arctic affinities from Fladbury, Worcestershire. *Quat. J. Geol. Soc. Lond.*, **118**, 103–23.

Coope, G. R. (1968). An insect fauna from Mid-Weichselian deposits at Brandon, Warwickshire. *Phil. Trans. Roy. Soc. Land.*, **B254**, 425–56.

Coope, G. R. (1970). Interpretations of Quaternary insect fossils. *Annual Reviews of Entomology*, **15**, 97–120.

Coope, G. R. and Brophy, J. A. (1972). Late Glacial environmental changes indicated by a coleopteran succession from North Wales. *Boreas*, **1**, 97–142.

Coope, G. R. (1973). Tibetan species of dung beetle from Late Pleistocene deposits in England. *Nature*, **245**, 335–6.

Coope, G. R. (1978). Constancy of insect species versus inconstancy of Quaternary environments. In *Diversity of insect faunas*, (ed. L. A. Mound and N. Waloff), pp. 176–87. Symposia of the Royal Entomological Society, Vol. 9. Blackwell, Oxford.

Coope, G. R., Shotton, F. W., and Strachan, I. (1961). A Late Pleistocene fauna and flora from Upton Warren, Worcestershire. *Phil. Trans. Roy. Soc. Lond.*, **B244**, 379–421.

Dansgaard, W., White, J. W. C., and Johnsen, S. J. (1989). The abrupt termination of the Younger Dryas event. *Nature*, **339**, 532–3.

d'Orchymont, A. (1927). Über zwei neue diluviale *Helophorus* Arten. *Sitzungsberichte und Abhandlungen der Naturwissenschaftlichen Gesellschaft Isis*, pp. 100–4. Dresden (1926).

Elias, S. A. (1994). *Quaternary insects and their environments*. Smithsonian Institute, Washington DC, pp. 1–284.

Hammond, P., Morgan, Anne, and Morgan, Alan (1979). On the *gibbulus* group of *Anotylus*, and fossil occurrences of *Anotylus gibbulus* (Staphylinidae). *Systematic Entomology*, **4**, 215–21.

Lowe, J. J. and Walker, M. J. C. (1984). *Reconstructing Quaternary environments*, pp. 1–389. Longman, London.

Matthews, J. V., Jr. (1970). Two new species of *Micropeplus* from western Alaska with remarks on evolution of the Micropepliinae (Coleoptera, Staphylinidae). *Can. J. Zool.*, **48**, 779–88.

Matthews, J. V., Jr. (1974). A preliminary list of the insect fossils from the Beaufort Formation, Meighen Island, District of Franklin. *Geol. Surv. Can. Papers*, **74–1A**, 203–6.

Matthews, J. V., Jr. (1976). Insect fossils from the Beaufort Formation: geological and biological significance. *Geol. Surv. Can. Papers*, **76–1B**, 217–27.

Osborne, P. J. (1974). A Late-glacial insect fauna from Lea Marston, Warwickshire. *Proc. Coventry & District Nat. Hist. & Sci. Soc.*, **4**, 209–13.

Ruddiman, W. F. and McIntyre (1981). The North Atlantic ocean during the last deglaciation. *Palaeogeog., Palaeoclimatol., Palaeoecol.*, **35**, 145–214.

Ruddiman, W. F. and Raymo, M. E. (1988). Northern Hemisphere climatic regimes during the past 3 Ma: possible tectonic connections. *Phil. Trans. Roy. Soc. Lond.*, **B318**, 411–30.

Shackleton, N. J. (1977). Oxygen isotope stratigraphy of the Middle Pleistocene. *British Quaternary studies: recent advances*, (ed. F. W. Shotton), pp. 1–16 Clarendon Press, Oxford.

Shotton, F. W. *et al.* (1993). The Middle Pleistocene deposits of Waverley Wood Pit, Warwickshire, England. *J. Quaternary Sci.*, **8**, 293–325.

Taylor, B. J. and Coope G. R. (1985). Anthropods in the Quaternary of East Anglia—their role as indicators of local palaeoenvironments and regional palaeoclimates. *Mod. Geol.*, **9**, 159–85.

Ullrich, W. G. and Coope, G. R. (1974). Occurrence of the east palaearctic beetle *Tachinus jacuticus* Poppius (Col. Staphylinidae) in deposits of the Last glacial period in England. *J. Entomol.*, **B42**, 207–12.

Ullrich, W. G. (1975). *Monographie der Gattung Tachinus Gravenhorst (Coleoptera, Staphylinidae), mit Bemerkungen zur Phylogenie und Verbreitung der Arten*, pp. 1–365. Dissertation, Christian-Albrechts Universität zu Kiel.

Walker, I. R. (1987). Chironomidae (Diptera) in Palaeoecology. *Quaternary Science Reviews*, **6**, 29–40.

Walker, I. R. and Mathews, R. W. (1988). Late Quaternary fossil Chironomidae (Diptera) from Hippa Lake, Queen Charlotte Islands, British Columbia, with special reference to *Corynocera* Zett. *Can. Entomol.*, **120**, 739–51.

Wilkinson, B. J. (1984). Interpretations of past environments from sub-fossil caddis larvae. *Proceedings of the 4th International Symposium on Trichoptera*, Series Entomologica, Vol. 30, pp. 447–52. Junk, The Hague.

Williams, N. E. (1988). The use of caddisflies (Trichoptera) in paleoecology. *Palaeogeog., Palaeoclimatol., Palaeoecol.*, **62**, 493–500.

5

Bird extinctions in the central Pacific

Stuart L. Pimm, Michael P. Moulton, and Lenora J. Justice

5.1 Introduction

The Pacific islands faunas and floras are well known for their high extinction rates (Diamond 1984*a,b*; Smith *et al.* 1993; Milberg and Tyrberg 1993; Humphries and Fisher 1994). Human colonization has driven these changes in the last few millennia. The first wave spread eastwards from the East Indies reaching Melanesia and Micronesia about 4000 ya (years ago), Fiji and Samoa 3500 ya, the Marquesas 2000 ya, and the outliers of Hawai'i, Easter Island, and New Zealand within the last 1500 years. European exploration started with Magellan and Mendaña in the 16th century and was all but complete when Cook died at Kealakekua, Hawai'i in 1779. Colonization started later: the first missionaries arrived in 'Tahiti in 1795 and Hawai'i in 1820, for example. Both waves brought alien species to the region although the peak period of introductions is much more recent. Most bird introductions to Hawai'i occurred since 1920.

Before humans arrived, Pacific islands housed a diverse avifauna. The first human colonizations caused extensive extinctions (Olson and James 1982). Estimating the true extent of the depredation is not trivial, however. The second wave of colonists also caused extinctions—and they continue to the present with many species surviving at population sizes too low for long-term persistence.

The purpose of this chapter is to document the depressing statistics. We restrict our analyses to the land birds of the central Pacific: the area between 30 °N and 30 °S and 140 °E and 100 °W. By land birds, we mean those species not tied to the ocean for feeding. We exclude terns, boobies, and shearwaters, for example, but include ducks, most herons (but exclude reef heron), and other birds of freshwater habitats. We start with the Hawaiian islands. They have been the most thoroughly explored and documented. We will discuss Hawai'i in detail because those details cause important and systematic biases to the extinction rates, raise difficult questions about the timing of extinctions, and point to the importance of introductions in further reducing endemism. We then consider the islands of south-eastern Polynesia and finally move westwards. The details will not be repeated, although they apply throughout.

5.2 The Hawaiian islands

The Hawaiian islands are of two types: from north to south, the six main islands (Kaua'i, O'ahu, Moloka'i, Maui, Lana'i, and Hawai'i) and their smaller attendants are relatively young, shield volcanoes in various stages of erosion. From every island, one can see at least one other island. To the north and west, the much smaller, older, and more isolated islands consist of coral atolls atop the eroded bases of once huge volcanoes. Only Laysan, Lisianski, and Nihoa supported land birds. On the main islands, the remains of birds in caves in lava tubes (and other such places) provide an unusually detailed record of the islands' faunas (Olson and James 1982, 1991; James and Olson 1991).

The not-so-noble savages

The total set of species consists of those that survived Polynesian contact, to be collected or observed by two centuries of biologists ('skins') and those that did not ('no skins'). A second dichotomy is between those species for which we have fossils ('bones') and those for which we do not ('no bones'). The combination of the two sets gives four subsets. Obviously, we are totally ignorant of those species lost during contact and for which we have no fossils ('no bones and no skins'). Other things being equal, we can estimate their number as

$$\frac{\text{'bones but no skins'} \times \text{'skins but no bones'}}{\text{'bones plus skins'}}$$

The analogy to the familiar mark–recapture method of estimating numbers is obvious. As an example, 29 species of non-passerine birds have been found only as fossils in Hawai'i, while 2 are known only from modern collections or observations, and 7 are known from both fossils and collections. The estimated number of missing species is 8 (to the nearest integer value.) The formula is sensitive to the errors that come with small samples and we apply it only when there are numbers sufficient to yield sensible estimates.

Nor do all species fossilize well. A bone from a large-bodied species may survive better than its fragile equivalent from a small-bodied species. Common species are more likely to be found than rare ones. Some species are more vulnerable to extinction—not only the rare ones, but also those that were the perfect accompaniment to *poi* (a dish made from taro root). We may never know how many rare small-bodied species have disappeared. Nonetheless, there are two ways in which we can refine our estimates. We can distinguish the large, likely to fossilize, good-for-the-calabash species (typically non-passerines) from the small species (typically passerines). In an independent classification, we can assume that we are more likely to find widespread species than those restricted to one island. Even within an island, those restricted to it may have been less numerous than those found more widely (see Lawton, Chapter 10). This is true of the species that still survive in the Hawaiian islands.

Finally, not all islands yield fossils equally. Kaua'i provides Olson and James with 1 site, O'ahu with 3, Moloka'i with 2, Maui with 11, and Lana'i and Hawai'i have no 'representative' sites (Olson and James 1991). We do not require mark–recapture studies to cover all the locations that house a population, only that the areas sampled are adequate given the mixing of the individuals. The same concern applies here. If the islands' species were well mixed, then this would not be a problem. As we shall see, most of the islands' species are limited to one island. They are far from being 'well mixed'.

Separating the species into non-passerines and passerines, leads to estimates of 8 and 21 missing species respectively. Dividing the passerines into those found on only one island and those found on several, yields estimates of 20 and 1 respectively. (That these add to 21—the estimated number for both passerine groups combined—is reassuring, but coincidental.) The number of missing non-passerines found on more than one island we estimate as 12 and there are too few data to estimate the number of non-passerines found on only one island. We conclude that there are 21 species of passerines and between 8 and 12 species of non-passerine bird that fell to the Polynesians without trace. We think these estimates are conservative.

Separating all species into those found on only one island and those found on several, leads to estimates of 46 and 3 respectively. The latter number is smaller than our estimate for the non-passerines alone. The former number would be increased almost in direct proportion by new finds of species known only from fossils and only from Lana'i or Hawai'i. Almost all (25 of 26) species of non-passerines found on only one island are found only as bones. Many were of flightless species that could not have occupied more than one island. Thus the number of missing non-passerines restricted to one island is likely to be large. On the four islands for which we have good fossil records, the estimated numbers of missing non-passerines on those islands equals or exceeds the number of missing passerines. The estimate of 8–12 missing non-passerines is surely too low: there are likely to be more missing species and they are likely to be restricted to just one island.

Missing skins and false accusations

The extinctions occurring within the last 200 years would seem to be easier to document. Since 1800, 18 species have gone extinct on the main islands and one more species and two endemic subspecies have gone extinct on Laysan. This is also an underestimate, because distinguishing Polynesian- from European-caused extinctions is not always easy. How many species survived the Polynesians yet disappeared before collectors could shoot them? Obviously, if only large numbers of specimens are collected for most species, we would conclude that we had not missed any. This is not the case.

The late-19th century saw the most intensive bird collections. By then, trade (including whaling and ranching), the introduced species it brought in general, and rats in particular, had been a feature of the islands for 60 years (Atkinson 1985). The collectors may have missed species that went extinct quickly for they even missed some that did not. One species, the *po'ouli*, was first described as recently as 1973. Twenty years later, it is so rare that in the last two years only one of several extensive and intensive surveys has found isolated individuals. It is likely to go extinct soon and it could well have disappeared without a trace. There are fossil *po'ouli*. Had there been no sightings, its loss would have been blamed on the Polynesians. The Lana'i hookbill is not known from fossils. Munro collected the only specimens in the 1920s and, until recently, they were assumed to be aberrant individuals of other species (Berger 1981; Pratt *et al.* 1987), despite Munro's field notes (Munro 1944). It, too, was almost lost. The rail from Lisianski (in the north-western chain), was seen but not collected. The Moloka'i rail is known only from its calls (Berger 1981). There is a fossil rail for Moloka'i. We do not know if the two rails are the same. There is a fossil *'o'o* from Maui, but only one, unconfirmed sight record. The 'spotted Hawaiian rail' may have been immature of the other rail on Hawai'i, which itself is known from only a 'few' specimens (Berger 1981). In this regard, it is not unlike the *kioea*, the *'ula-'ai-hawane* (with only 5 specimens), and the greater *'amakihi* (with about 10) (Berger 1981). Elsewhere in the Pacific there is the mysterious starling (1 specimen), the Tahiti rail (no specimen, but illustrated following Cook's voyage), and the Tahitian sandpiper (3 specimens) (Pratt *et al.* 1987; Steadman 1989).

Given these near misses, it is likely that there are several species missing from the recent record. While it is likely that the Polynesians did exterminate most of the 43 species we know only as fossils, and the roughly 30–50 we estimate are missing, they may not have killed them all. Attributing 19th-century extinctions to the Polynesians is doubly unfair. First, it places false blame. Second, it causes us to underestimate the number of species we count as both fossils and skins and to overestimate the number of species found only as fossils. Both effects inflate our estimates of the number of missing species and we attribute their loss to the Polynesians. Of course, the species may not be found as fossils and their loss from the 'skins only' category would reduce the estimate of missing species.

Taxonomy and other modern problems

A further set of problems comes from considering the species that are still present in the islands. There are still unresolved taxonomic issues. As these are resolved by both conventional techniques and newer DNA-based methods (e.g., Tarr and Fleischer 1993), the number of species known from the islands tends to increase as local, rare forms are found to be distinct species. Moreover, there are many species whose continued presence on the islands is uncertain. Even if they do persist, some are so rare that their immediate extinction is likely in the absence of active demographic intervention. The details of the necessary taxonomic book keeping appear in the companion article to this one (Pimm *et al.* 1994) and only our summary appears below.

Haole moa—*foreign birds*

How Hawai'i acquired an introduced avifauna larger than any other area is a story we have told before (Moulton and Lockwood 1992; Moulton and Pimm 1985, 1987). Here, we will mention only the consequences to β-diversity, and use percentage endemism as its measure. The original land birds of Hawai'i contain only three species (*pueo*, moorhen, night heron) that are not endemic to the islands. All the passerines are endemic. Polynesian extinctions and introductions reduced non-passerine endemism from 93% to 67%, and recent extinctions to

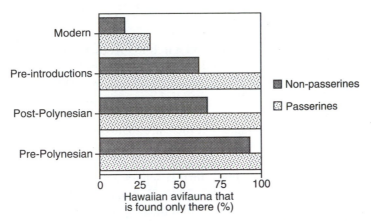

Fig. 5.1 In Hawai'i, the percentage of endemic passerines and non-passerines was high before human contact. The two waves of colonization caused extinctions, but these reduced percentage endemism relatively little. The major impacts on endemism were the species brought by the Polynesians and especially those brought in the last two centuries.

62%. Yet modern species introductions reduced non-passerine endemism to a mere 16%. Currently, only about one-third of the passerine birds are endemic. Almost all of the birds (species and individuals) seen below 1000 m are now alien (Fig. 5.1).

Summary

Eleven species are abundant enough to not generate immediate concern about their future. Twelve species are endangered, a further 12 are so rare that they are unlikely to survive in the wild (even if they still exist), and 18 are extinct. Forty-three species are known only from fossils, and we estimate that between 30 and 50 species have been lost without a trace. Of the roughly 125 to 145 species that once inhabited the main Hawaiian islands, 90 to 110 are extinct and based on recent experience 100 to 120 will be the figure by the end of the decade. Less than 10% of the species have really survived human impact. Figure 5.2 provides a summary of these percentages. Humans brought most of the species now present in the islands.

5.3 South-eastern Polynesia

South-eastern Polynesia is a loose group of island clusters, consisting of the Cooks, Societies, the Tuamotus, the Marquesas, and the remote islands of Henderson and Pitcairn. We justify their joint consideration because only 6 of the more than 60 species are found outside the area. Most of the islands are atolls, but some are the relatively large remains of old volcanoes. They also introduce the third island type, Makatea islands, of which Makatea in the Tuamotus is the type specimen. These are high islands of uplifted corals and have limestone substrates.

(Islands can be mixtures; the north of Guam in the Marianas is of this type, the south is volcanic.)

Steadman (1989, 1991; Steadman and Kirch 1990) describes fossil faunas from Henderson, Huahine (in the Societies), three islands in the Cooks, and four islands in the Marquesas. For all species combined, we estimate the missing species to be 0, 11, 4, and 6, respectively. Of course, some of these 21 species might survive elsewhere in the group because the islands share many of their species. For the entire region, we estimate that there are 16 missing non-passerines and 4 missing passerines. We also estimate that for both groups combined, there are 22 missing species that occurred on only one island group, and 1 species missing across the entire region. We cannot further separate data.

The range of missing species is thus 20 to 23. A further 20 species are known only as fossils. Forty-seven species survived the Polynesians, suggesting they may have exterminated about 45% of the fauna. Again, while estimates of missing species increase the Polynesian impact, we simply do not know how many of these species fell, for example, to the cats deliberately introduced to 'Tahiti on the expedition that carried Banks. Of the 47 survivors, only 8 species are known to have become extinct since European exploration. Four more species are endangered, three occur only on uninhabited islands, and three more have such low numbers that their existence is uncertain (Pratt et al. 1987; Holyoak 1980). About one-third of the bird species have survived human impact. Figure 5.2 summarizes these changes.

These islands have introduced species too. As always, the Polynesians brought moa (jungle fowl, *Gallus gallus*) and more recent efforts brought 41 passerine species to 'Tahiti of which 7 survived (Lockwood et al. 1993). There are three other non-passerines introductions.

5.4 The Marianas

The largest islands in the Marianas are of the Makatea type. Their numerous caves were defended vigorously during the Second World War and those on Guam, Tinian, and Saipan were badly damaged in the fighting. Rota, in contrast, was not contested, and Steadman has explored its caves (Steadman 1992). Seven species are found only as fossils on Rota, and there is a further species found only on Tinian. Of these eight species, two are known from elsewhere (Micronesian pigeon, purple swamphen), two are of uncertain affiliation and may be of species found elsewhere in the islands. Four species are thought to be new ones, including a flightless duck and a giant ground dove.

We estimate that five non-passerines and two passerines are missing from the fossil record. These numbers would be increased by finds of species restricted to one island, but several of Steadman's finds show that species were once more extensively distributed. Twenty species survived to recent times. Of the 28 known species, 14 are endemic and 3 more are distinctive endemic subspecies. Of the 20 survivors to modern times, 5 are extinct, 1, the Guam rail, occurs only in captivity although attempts are being made to re-introduce it to the wild, Witteman et al.

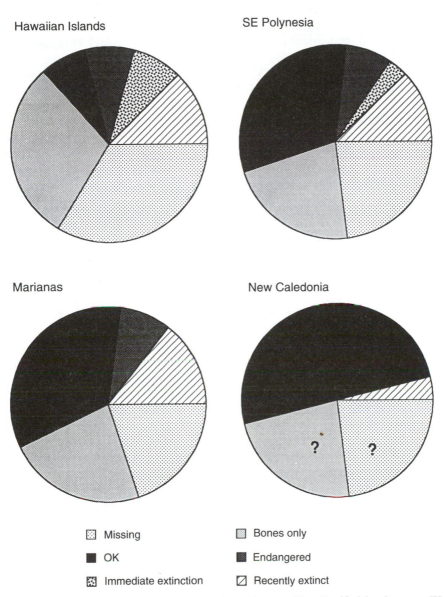

Fig. 5.2 The proportional composition of the avifaunas of four Pacific island groups. We recognize six categories. Species that are sufficiently abundant to be in no immediate danger of extinction ('OK'), those that are likely to go extinct, but not immediately ('endangered'), those that are likely to go extinct immediately—and which may be extinct already ('immediate extinction'), those that have become extinct in historical times ('extinct'), those known only as fossils ('bones only'), and those we estimate are missing from the fossil record ('missing'). For New Caledonia, the numbers of 'bones only' and 'missing' are crudely estimated as twice the numbers known and estimated for the non-passerines alone. (The data on fossil passerines are not yet available.)

1990), and three species are endangered. Only the Guam rail, the Guam kingfisher (possibly only a subspecies, although a distinctive one), the Guam flycatcher, and the taxonomically debatable Marianas mallard have gone from the wild.

In summary, of an estimated 35 species, 15 fell to the islands' first colonists, 5 more since then, but 12 have survived human occupation—at least until now (see Fig. 5.2). Consider the island of Guam: it has only one species that may be considered safe, two are endangered, one more (the Marianas crow) survives but rarely breeds, and the island has lost a dozen species. The cause of these extinctions—the introduced brown tree snake—is well known (Savidge 1987; Pimm 1987). So, too, is the snake's spread to Saipan and the occurrence of dead snakes in Hawai'i. The potential consequences to the remaining Pacific faunas are obvious.

5.5   The western Pacific

Finally, we turn to the faunas of New Caledonia, Vanuatu, and the group of islands than contains the Fiji, Samoa, and Tonga. There are many atolls in this cluster, but it also contains many large islands, in fact, islands larger than any we have discussed so far. We start with Fiji and our personal, identical yet independent impression of its largest island, Viti Levu.

Our first impressions of the islands examined so far were not favourable. Guam has no forest birds and lowland Hawai'i has only a few rare and local native forest birds. To see native Hawaiian birds, one gets cold, wet, and tired. On Viti Levu, in contrast, one can see native birds from one's beach front condominium. Indeed, only 4 of Viti Levu's 47 land birds occur only in the rainforest, and 17 occur commonly in agricultural and suburban habitats (Watling 1982). Viti Levu has introduced species (nine of them), but they do not comprise the majority of the islands individuals or species. This impression of paradise continues as one studies Watling's book on the birds of Fiji, Tonga, and Samoa. There are 67 land bird species in the region, and 46 are confined to it. Of these, only 4 have become extinct in historical times, only 4 more are considered 'rare'. One species is endangered (the *niuafo'ou* megapode), in part because it occurs on only one island and in part because of its peculiar habits. (It incubates its eggs in hot, volcanic ash.) One further species (the long-legged warbler) was once considered rare, but Watling considers it merely hard to find.

Fossil data are few. Steadman has explored 'Eua of the Tonga group. Our analyses of this single island, predicts only one missing species. Three more species are known only from bones, there are three extinctions in historical times, but all occur widely throughout the region and 12 species survive until the present.

This scarcity of recent extinctions extends to the Vanuatu chain, where Bregulla (1992) reports no recent extinctions in his discussion of the islands' 55 species. He reports only one species near extinction and it is widespread elsewhere. There are two species that are restricted to montane forest on just one island each and both are common in their very restricted ranges. (There are about 30 islands or island groups in the Vanuatu chain, stretching over 1000 km from north to south.)

New Caledonia has 56 species (31 non-passerines) of birds known historically, of which 4, all non-passerines, have become extinct in recent times (Hannecart and Letocard 1980). So far, only fossil non-passerines have been described (Balouet and Olson 1989). These add 12 species known only as fossils, and we estimate that there are another 12 missing. This island's species include an endemic family (the Kagu, Rhynochetidae) with 2 species, 1 fossil and 1 modern, and 1 fossil species of uncertain familial affinities, the giant, flightless *Sylviornis neocaledoniae*. Since the numbers of non-passerines are roughly the same, an even rougher guess for the number of lost and fossil passerines might be the same as for the non-passerines (i.e., 24). This would bring New Caledonia's total to 104 species; Fig. 5.2 is a summary.

5.6   Paradise or slaughterhouse?

Something is very wrong in this paradise. First, recall our estimates of pre-colonization species richness: Hawai'i, 125–145 species, south-eastern Polynesia 87–90 species, the Marianas 35 species, Fiji, Tonga, and Samoa, 67 plus extinctions, New Caledonia 104, and Vanuatu 55 plus extinctions. Exclude the Marianas, for they are low and dry islands; the rest have both small atolls and large, high, volcanic islands. Should we conclude (approximately) that islands have *more* bird species on them the *smaller* they are and the *further* they are from source areas for potential colonists? No, we do not actually believe that a quarter of a century of island biogeography has got the patterns exactly the wrong way around.

When all the fossils are found, we predict that the actual species-richness of the Fiji, Tonga, and Samoa group should be at least that of the Hawaiian islands: the 'plus extinctions' should exceed the 67 known species. The Hawaiian islands are more remote and they are closer to each other, so allowing less between-island diversity. It would be a considerable biogeographic anomaly if the Fiji, Tonga, and Samoa group did not have more species than the smaller, more distant group of islands in south-eastern Polynesia! First-wave colonizations seem to remove about 50% of an archipelago's fauna. A reasonable prediction would be that the Fiji, Tonga, and Samoa group lost 60 or more species.

Similarly, we predict that the missing fossils of Vanuatu will also roughly double its numbers of species. This would be broadly comparable with the situation in nearby New Caledonia as well as the other islands.

Less confidently and more controversially, we suggest that even these numbers (roughly 50 species for each of New Caledonia and Vanuatu) are underestimates. These numbers lead to an almost flat distribution of species richness with increasing distance from sources of colonists across the Pacific. This may be correct. For remote islands speciation, not colonization is the principal factor increasing species richness. Nonetheless, the increased colonization rates we

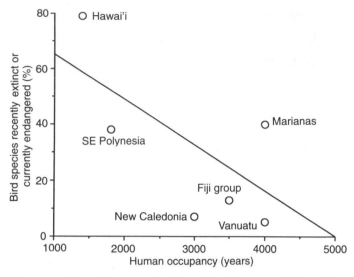

Fig. 5.3 The proportion of the recent avifauna that is now extinct, endangered, or in immediate danger of extinction drops as the age of human colonization increases. This suggests that most human-sensitive species have already disappeared from the islands where human occupation is the longest. The Marianas have an unusually high numbers of modern losses. The recent colonization of the islands by the brown tree snake is the cause (see text).

expect for the islands in the western Pacific, not only provide more species directly (their faunas have less endemics than those in the east and north), but these colonists may seed the process of speciation. Speciation on New Caledonia was sufficiently vigorous to produce at least one endemic family (the Kagu) and the bizarre *Sylviornis*.

Another reason for considering the western Pacific estimates of extinction to be on the low side comes from considering the proportion of modern species that are endangered or known to have gone extinct in the last two centuries. This proportion is high in Hawai'i and drops towards the west, with increasing age of human settlement (Fig. 5.3).

Perhaps the islands in the western Pacific lost the 'wimps'—the species that were in any way sensitive to human activities—a long time ago. This would explain why so many Fijian species occur around human disturbance and so few in the forest and why the pattern is reversed in Hawai'i. Fiji may have lost its forest specialists a long time ago. We would explain Europe's lack of rare birds the same way: the 'wimps' are long gone. (Jared Diamond, pers. comm., suggests that elsewhere, frequent cyclones preclude forest specialists, because the forest is so frequently disturbed.) The consequence of this explanation is that New Caledonia, Vanuatu, and the Fiji, Tonga, and Samoa group should have lost even more than the half of their fauna typical of more recently colonized islands. These recently colonized islands are still losing the 'wimps'.

Will the missing species be found? We are not sure. As one moves westwards across the Pacific, humans arrived and birds were lost at earlier dates. Not only might intact bones become scarcer, but their hiding places may be more likely to be destroyed. It is possible, of course, that the numbers of species may be determined by local factors, such as island height and isolation, the range of habitats, the presence of land crabs (Atkinson 1985), and many other details. If these factors are really important in determining diversity, their effects would not lead to a predictable decline in species-richness eastwards across the Pacific.

5.7   Conclusions

The number of species lost when humans colonized Pacific islands is approximately double the number of known fossil species and involved about half the species present. Estimated rates of extinction and current endangerment drop with the period of human occupancy, suggesting that those areas colonized first have lost most of their sensitive species.

While adding further crimes to the first colonists, we point to the considerable uncertainties about which species survived to the modern era. In Hawai'i, for example, intensive collecting followed European contact by half a century. This period is certainly long enough for many species to have been lost except as fossils. Indeed, several species were not collected and there were several 'near misses'.

Extinction rates for the last 200 years frequently ignore a substantial number of species whose numbers are too small to survive without immediate active human intervention. Even more species are endangered, meaning that their long-term survival is unlikely.

Faunal changes have done more than reduce local diversity. Single-island endemics predominate among the species that are lost. There are several explanations for this. Obviously, they would be more vulnerable than species that occur on several islands, even if the per island extinction probabilities were similar for all species. Their probabilities may be higher, because such species may adopt life histories that lead them to being vulnerable (flightlessness, for example). (Their per island extinction rates may appear higher spuriously, because all species on the way to extinction must eventually be found on only one island.) Whatever the reasons, their extinction reduces β-diversity: the percentage of endemic species drops. Even more dramatic in its effects on endemism is the introduction of alien species. Introductions reduce endemism to levels where, in Hawai'i for example, the great majority of the species are not endemic.

Summary

The first wave of human colonists spread across the Pacific from 4000 to 1000 years ago. That they caused many extinctions is well known from fossil finds. We estimate how many fossil species were missed—the answer is roughly half—and so estimate the true extinction rate. The first colonists exterminated roughly half the species on each island group. Some of these estimated extinctions are falsely attributed to the first colonists, because intensive collection often began a half century after the damage initiated by European discovery. Even taken at face value, these recent extinctions are too few. Many species are so critically endangered that we know neither whether they still survive nor how to save them. Interestingly, there are fewer recent extinctions and currently endangered species in the islands of the western Pacific—those occupied first by humans. We suggest that the species sensitive to human occupation died out long ago in these areas. If so, these islands would have lost even more than half of their bird species.

Acknowledgements

SLP thanks the Pew Charitable Trusts and the Nature Conservancy of Hawai'i for financial support, Becky Ashe for compiling data, his wife Julia for continued and enthusiastic support in running his laboratory, Jared Diamond for comments, and Eric Greene, Stephen Killeffer, and Teresa Weronko for good company in the cold and wet of upland Maui.

References

Atkinson, I. A. E. (1985). The spread of commensal species of *Rattus* to oceanic islands and their effects on island avifauna. In *Conservation of island birds*, Technical Publication No. 3. (ed. P. J. Moors), pp. 35–81. International Council for Bird Preservation, Cambridge.

Balouet, J. C. and Olson, S. L. (1989). Fossil birds from late Quaternary deposits in New Caledonia. *Smithsonian Contributions to Zoology*, **469**. Smithsonian Institution Press, Washington, DC.

Berger, A. J. (1981). *Hawaiian birdlife*, (2nd edn). University of Hawai'i Press, Honolulu.

Bregulla, H. L. (1992). *Birds of Vanuatu*. Nelson, Oswestry.

Diamond, J. M. (1984a). Historic extinctions: a rosetta stone for understanding prehistoric extinctions. In *Quaternary extinctions: a prehistoric revolution*, (ed. P. S. Martin and R. G. Klein), pp. 824–62. University of Arizona Press, Tucson.

Diamond, J. M. (1984b). 'Normal' extinctions of isolated populations. In *Extinctions*, (ed. M. H. Nitecki), pp. 191–246. University of Chicago Press.

Hannecart, F. and Letocart, Y. (1980). *Oiseaux de Nouvelle Caledonie et des Loyauteés*. Clark and Matheson, Auckland.

Holyoak, D. T. (1980). *Guide to Cook Islands birds*. Privately printed.

Humphries, C. J. and Fisher, C. T. (1994). The loss of Banks's legacy. *Phil. Trans. Roy. Soc.*, **B343**, 3–9.

James, H. F. and Olson, S. L. (1991). Descriptions of thirty-two new species of birds from the Hawaiian islands: Part II. Passeriformes. *Ornithological Monographs*, **46**. American Ornithologists' Union, Washington, DC.

Lockwood, J. L., Moulton, M. P. and Anderson, S. K. (1993). Morphological assortment and the assembly of introduced passeriforms on oceanic islands: Tahiti versus O'ahu. *Amer. Nat.*, **141**, 398–408.

Milberg, P. and Tyrberg, T. (1993). Naive birds and noble savages—a review of man-caused prehistoric extinctions of island birds. *Ecography*, **16**, 229–50.

Munro, G. C. (1944). *Birds of Hawaii*. Bridgeway Press, Rutland, Vermont.

Moulton, M. P. and Lockwood, J. L. (1992). Morphological dispersion of introduced Hawaiian finches: evidence for competition and a Narcissus effect. *Evol. Ecol.*, **6**, 45–55.

Moulton, M. P. and Pimm, S. L. (1985). The extent of competition in shaping an experimental avifauna. In *Community ecology*, (ed. J. Diamond and T. Case), pp. 80–97. Harper & Row, New York.

Moulton, M. P. and Pimm, S. L. (1987). Morphological assortment in introduced Hawaiian passerines. *Evol. Ecol.*, **1**, 113–24.

Olson, S. L. and James, H. F. (1982). Fossil birds from the Hawaiian Islands: evidence for wholesale extinction by man before western contact. *Science*, **217**, 633–5.

Olson, S. L. and James, H. F. (1991). Descriptions of thirty-two new species of birds from the Hawaiian islands: Part I. Non-passeriformes. *Ornithological Monographs*, **45**. American Ornithologists' Union, Washington, DC.

Pimm, S. L. (1987). The snake that ate Guam. *Trends Ecol. Evol.*, **2**, 293–5.

Pimm, S. L., Moulton, M. P. and Justice, L. J. (1994). Bird extinctions in the Central Pacific. *Phil. Trans. Roy. Soc. Lond.*, **B344**, 27–33.

Pratt, H. D., Bruner, P. L. and Berrett, D. G. (1987). *A field guide to the birds of Hawai'i and the tropical Pacific*. Princeton University Press, Princeton, NJ.

Savidge, J. A. (1987). Extinction of an island avifauna by an introduced snake. *Ecology*, **68**, 660–8.

Smith, D. M., May, R. M., Pellew, R., Johnson, T. H. and Walter, K. S. (1993). Estimating extinction rates. *Nature*, **364**, 494–96.

Steadman, D. W. (1989). Extinction of birds in eastern Polynesia: a review of the record, and comparisons with other Pacific island groups. *J. Arch. Sci.*, **16**, 177–205.

Steadman, D. W. (1991). Extinct and extirpated birds from Aitutaki and Atiu, southern Cook Islands. *Pacific Sci.*, **45**, 325–47.

Steadman, D. W. (1992). Extinct and extirpated birds from Rota, Mariana Islands. *Micronesica*, **25**, 71–84.

Steadman, D. W. and Kirch, P. V. (1990). Prehistoric extinction of birds on Mangaia, Cook Islands, Polynesia. *Proc. Nat. Acad. Sci. USA*, **87**, 9605–9.

Tarr, C. L. and Fleischer, R. (1993). Mitochondrial DNA variation and evolutionary relationships in the 'Amakihi complex. *Auk*, **110**, 825–31.

Watling, D. (1982). *Birds of Fiji, Tonga, and Samoa*. Millwood Press, Wellington.

Witteman, G. J., Beck, R. E., Pimm, S. L. and Derrickson, S. (1990). The extinction and re-introduction of the Guam Rail. *Endangered Species Update*, **8**, 36–9.

6

Extinctions in Mediterranean areas

Werner Greuter

6.1 Introduction

The Mediterranean area is known as a region of high floristic diversity. When one delimits the area, as was done for the purposes of the *Med-Checklist* (Greuter *et al.* 1984–9), to include all countries bordering the Mediterranean Sea plus Portugal, Bulgaria, the Crimea, and Jordan, its wild flora of *c.* 24 000 species comes close to one-tenth of the quarter million known vascular plant species of the globe (Greuter 1991). Jointly, the five areas of the world with a Mediterranean-type climate (the Mediterranean proper, California, central Chile, the Cape region, and south-western Australia) 'total some 70 000 or more species and represent a major part of the higher plant diversity of the world, exceeding the combined floras of tropical Africa and Asia' (Heywood 1993).

As I have demonstrated for the Mediterranean area proper (Greuter 1991), this richness is not caused by high local species density (on average, the species numbers in small areas are not higher than in central Europe) but by small mean distributional ranges, reflected by a remarkable number of narrow endemics. Many of these are extremely rare, occurring in low numbers and limited to a single or few localities. It is reasonable to assume that they are threatened, and the question of whether and to what extent their vulnerability has already resulted in extinction is certainly appropriate.

6.2 Recorded extinctions in the Mediterranean area

For the purpose of this chapter, the Mediterranean area will be equated with the area covered by the *Med-Checklist* (Greuter *et al.* 1984–9). Phytogeographically, this is at best a rough approximation, since that range extends beyond the Mediterranean summer drought domain to the north and south while not covering it totally toward the east. Yet for practical purposes accepting the *Med-Checklist* limits makes sense, because within them complete, updated figures are available for those higher plants already investigated (i.e., for *c.* 45% of the total vascular flora). I shall restrict this analysis to vascular plant taxa in the ranks of species and subspecies; I will not consider partial extinctions, concerning only a portion of a taxon's range, but only total extinction.

Information concerning extinctions is in need of constant updating. The general inventories of threatened and endemic plants of the area (Anon. 1983 for Europe; Lucas 1980 for South-Western Asia and North Africa) are already badly out of date, as is my own partial survey of extinct taxa (Greuter 1991) for the families so far treated in the *Med-Checklist*. A number of country Red Data Books (or floristic inventories including red data categories) has been forthcoming in recent years, in which the field-based knowledge of local botanists has been incorporated: Conti *et al.* (1992) for Italy, Ekim *et al.* (1989) for Turkey, Gamisans and Jeanmonod (1993) for Corsica, Gómez-Campo (1987) for Spain and the Balearic Islands, Hadidi *et al.* (1992) for Egypt, Lanfranco (1989) for Malta, Velčev *et al.* (1984) for Bulgaria, and Wraber and Skoberne (1989) for Slovenia. All these, and a few other sources, have been used to re-write the inventory of known Mediterranean plant extinctions as given in Table 6.1. Underlining the needs of continuous updating, the last four entries in Table 6.1 were added in press, as a result of the recently published inventory of the threatened plants of Greece by Phitos *et al.* (1993).

Several taxa had to be added to those listed as extinct in previous overviews, but others could be omitted for a variety of reasons. Some have been found to still survive, others are 'ghost species' that have never existed or have been sunk into synonymy, and a few were the likely product of occasional hybridization. There is also one (mainly central European) category of plants that I decided not to consider: semi-domesticated weeds of arable fields, because for several it is impossible to be sure whether and where they still exist in the wild, whereas most if not all are kept alive in germplasm collections. Three brome-grass species as well as *Filago neglecta*, *Lolium remotum*, and *Silene linicola* belong to this category.

The total number of Mediterranean higher plants presumed to be extinct is thus 37, of which 35 were species and 2, subspecies. This compares to a total of about 23 300 native Mediterranean plant species, or 29 000 taxa (species plus subspecies). The present Mediterranean extinction rate for vascular plants is therefore 0.15% at species level, or a mere 0.13% of the taxa.

6.3 Extinction records in other areas with a Mediterranean climate

No relevant information has been found for the Mediterranean portion of Chile, nor indeed for that whole country, but there are thorough recent synopses with comparable data on the extinct and threatened higher plants for the three other summer drought areas.

The Cape floristic region of South Africa, or fynbos biome, is the South African counterpart of the Mediterranean domain. It is famous for its floristic richness, with no less than 8580 native vascular plant species growing in a relatively small area (Bond and Goldblatt 1984). Twenty-six of them are presumed to be extinct, twice as many as for the whole remainder of southern Africa (Hall and Veldhuis 1985), which corresponds to an extinction rate of 0.3%. For southern Africa as a whole that rate is considerably lower (0.19%).

Table 6.1 Mediterranean species and subspecies of vascular plants presumed to be extinct in the wild

Taxon	Country	Family	References
? *Allium rouyi* Gaut.	Spain	Liliaceae	1,3
Astragalus pseudocylindraceus Bornm.	Anatolia	Leguminosae	2
Avenula hackelii (Henriq.) Holub	Portugal	Gramineae	1
Campanula oligosperma Damboldt	Anatolia	Campanulaceae	2
Cephalaria kesrouanica Mouterde	Lebanon	Dipsacaceae	5
*Coincya pseuderucastrum subsp. *puberula* (Pau) Valdés	Spain	Cruciferae	5
Cynoglossum foliosum (Paine) Greuter & Burdet	Jordan	Boraginaceae	5
Dianthus multinervis Vis.	Former Yugoslavia	Caryophyllaceae	5
*Diplotaxis siettiana Maire	Spain	Cruciferae	3, 4, 5
Fagonia taeckholmiana Hadidi	Egypt	Zygophyllaceae	6, 8
Fibigia heterophylla Rech. f.	Syria	Cruciferae	5
? *Geocaryum bornmuelleri* (H. Wolff) Engstrand	Greece	Umbelliferae	1
? *Geocaryum divaricatum* (Boiss. & Orph.) Engstrand	Greece	Umbelliferae	1
? *Hypericum setiferum* Stef.	Bulgaria	Guttiferae	9
Limonium dubyi (Gren. & Godr.) Kuntze	France	Plumbaginaceae	7
*Lysimachia minoricensis Rodr.	Balearic I.	Primulaceae	1, 3, 5
Morina subinermis Boiss.	Anatolia	Morinaceae	5
Onosma affinis Riedl	Anatolia	Boraginaceae	2
Onosma discedens Bornm.	Anatolia	Boraginaceae	2
Salvia peyronii Post	Lebanon	Labiatae	5
Sedum polystriatum R. T. Clausen	Anatolia	Crassulaceae	2
Silene oligotricha Hub.-Mor.	Anatolia	Caryophyllaceae	2
Silene rothmaleri P. Silva	Portugal	Caryophyllaceae	5
Silene tomentosa Otth	Gibraltar	Caryophyllaceae	5
Tephrosia kassasii Boulos	Egypt	Leguminosae	8
Teucrium leucophyllum Benth.	Anatolia	Labiatae	2
Thalictrum simplex subsp. *gallicum* (Rouy & Foucaud) Tutin	France	Ranunculaceae	7
Thymus oehmianus Ronniger & Soska	Former Yugoslavia	Labiatae	1
Trifolium acutiflorum Murb.	Morocco	Leguminosae	5
*Tulipa sprengeri Baker	Anatolia	Liliaceae	2
Verbascum calycosum Hausskn. & Murb.	Anatolia	Scrophulariaceae	2
? *Veronica euxina* Turrill	Bulgaria	Scrophulariaceae	1, 9
Viola cryana Gillot	France	Violaceae	1, 7
Alkanna sartoriana Boiss. & Heldr.	Greece	Boraginaceae	10
Centaurea tuntasia Halácsy	Greece	Compositae	10
Paronychia bornmuelleri Chaudhri	Greece	Caryophyllaceae	10
Polygala subuniflora Boiss. & Heldr.	Greece	Polygalaceae	10

An asterisk (*) denotes taxa still extant in cultivation and/or seed banks; a question mark (?) indicates doubt as to actual extinction. References: 1, Anon. (1983); 2, Ekim *et al.* (1989); 3, Gómez-Campo (1987); 4, Gómez-Campo (1990); 5, Greuter (1991); 6, Hadidi *et al.* (1992); 7, Lesouef and Olivier (1989); 8, Lucas (1980); 9, Velčev *et al.* (1984), 10, Phitos (1993).

The state of California is not quite co-extensive with the Mediterranean domain in North America, which extends beyond its border both in a northerly and southerly direction, but it certainly represents its core. It has a native flora of 5867 species, or 7036 taxa (species, subspecies, and varieties), according to Hickman (1993). Among these, 34 are presumed extinct in California of which 6 survive outside that state, the genuine extinctions adding up to 24 species, 3 subspecies, and 1 variety (Smith and York 1984). The extinction rate is thus 0.41% for species, or 0.4% for taxa, which is exactly the same as that calculated by Ayensu and DeFilipps (1978) for the whole of the United States.

The Mediterranean climate area of Australia is not easy to define. A summer drought regime extends right across the southern part of the continent, yet the floristically rich area that one considers as typically Mediterranean coincides with the south-western floristic province of Western Australia. The state of Western Australia has a vascular flora of 7465 native species (8300 taxa, excluding formae) of which 54 (55 taxa) are presumed extinct, roughly two-thirds of the 82 (83) recorded extinctions of the whole continent (Leigh and Briggs 1992). The species extinction rate of 0.72% (0.66% for taxa) is the highest mentioned so far, but when only the south-western province is considered it may rise as high as 1% (50 species out of an estimated total of 5000). The species extinction rate for the Australian continent with its 17 590 native species (Hnatiuk 1990) is 0.47%.

6.4　Extinction versus threat—a disquieting perspective

When one thinks of the alarming figures about ongoing global mass extinction that one reads in the newspapers, one may feel pleasantly surprised, or at least relieved, at the low rates of recorded extinctions in the floristically rich Mediterranean-climate areas. But actual loss is one thing, impending threat of loss another.

Let us, for once, set aside the questions of how real the threats faced by surviving plant species are, how reliable the assessments of these threats, and how comparable the assessment criteria in various parts of the world; and let us just compare the figures one finds in the literature (Table 6.2). Those for the Mediterranean proper are put together from three independent sources (Lucas 1980; Anon. 1983; Ekim et al. 1989) and relate to species and subspecies; those for the Cape floristic region (fynbos biome), from Hall and Veldhuis (1985), concern species alone; and those for California and Western Australia (Smith and York 1984; Leigh and Briggs 1992) encompass species, subspecies, and varieties.

This survey shows that the percentages of threat are not significantly different between these four widely distant areas, ranging from 10.2% to 17.5% (which tends to demonstrate that threat assessment has at least used comparable criteria). It also shows that, if the published figures are basically realistic, the impending threat is far worse than what has been put on record in matters of extinction. Threat rate and extinction rate differ by a factor varying between about 25 (Western Australia) and 125 (the Mediterranean proper). In other

Table 6.2 Extinction and threat rates in Mediterranean-type floras, in per mille and per cent of the native vascular plant taxa, respectively. Sources and specifications are given in the text

Area	Native flora (no. of taxa)	Extinct taxa (no.)	Extinction rate (‰)	Threatened taxa (no.)	Threat rate (%)
Mediterranean (*Med-Checklist*)	29 000	37	1.3	4251	14.7
Cape Province (fynbos biome)	8580	26	3.0	1300	15.2
California	7036	28	4.0	718	10.2
Western Australia	8300	55	6.6	1451	17.5

words, if the present losses may still be felt to be tolerable, future extinctions threaten to happen on a cataclysmic scale.

6.5 How reliable are our data?

Apart from a few exceptional, fully documented cases extinctions are surrounded by considerable uncertainty. Many of those that have been put on record may in fact not have happened as yet, and instead many may already have taken place unnoticed.

In Mediterranean-type areas few species are so well studied and so closely monitored that one may confidently expect their disappearance to be noted and recorded as soon as it happens. Many species have not been re-collected since they were first discovered, perhaps over a century ago, for the simple good reason that the place where they grow has never again been visited by a botanist; or if someone did visit it, it may have been in the wrong season when only seeds or underground storage organs were present. Frequently, the collecting localities are not exactly known, or have been misstated. It would therefore be unreasonable to assume extinction for a taxon that has not been collected recently unless there are other facts to support that assumption, such as a thorough but vain search of the known localities or obvious disturbance or destruction of the areas and habitats of former occurrence.

One must thus reckon with an unknown number of unrecorded extinctions, especially for areas with a disproportionately low avowed extinction rate. Think of the Maghreb countries of North Africa with their high number of endemics of which but a single one, *Trifolium acutiflorum* Murb. from Morocco, is presumed extinct.

Conversely—as is indicated by the qualification of recorded extinctions as 'presumed'—botanists have learnt to be cautious when asserting that a species has died out. The number of presumed extinctions recorded for the California flora has decreased from 44 to 34 within four years: the rediscovery of 11 taxa and the dismissal of 3 that had in fact never existed having more than outweighed the

4 newly recorded losses. As Smith and York (1984) have clearly demonstrated: 'a number of List 1A [presumed extinct] plants do not stay there for long once special attention is drawn to them and renewed efforts are made to rediscover them.'

For Greece alone, no less than three species reported missing have been rediscovered in the last few years. *Adonis cyllenea* Boiss. *et al.*, not re-found on Mount Killíni since 1854, was collected exactly 130 years later on nearby Mount Parniás where a few small populations survive (Strid 1986, 1992). It has now been brought into cultivation and is said to show promise as an ornamental garden plant. The fate of *Helichrysum taenari* Rothm., discovered in southern Pelopónnisos by Rothmaler in 1942, was uncertain for many years. No specimens existed: they were all burnt in the Berlin fire of 1943, and Rothmaler's (1944) original description was based on his memory and field-book notes. A number of experienced field botanists searched for it in vain. Yet Strid (1986, 1992) found it growing safely in its single locality in 1984, and in a second nearby location in 1991, and he succeeded in growing it as an attractive, if not fully hardy plant for the rockery. Finally, there is the case of *Onobrychis aliacmonia* Rech. f., as certain an extinction as one might think of since its only locality and its surroundings were flooded about 1975, together with large parts of the Aliákmon valley, due to the construction of a dam. Yet, it was rediscovered in 1985 in a completely different environment and at a distance of 400 kilometres, in a Pelopónnisos locality (Greuter 1987), where meanwhile its population has again dramatically declined due to human action (G. Iatrou, pers. comm.). Is it due to disappear for a second time and, if so, may we hope for yet another rediscovery elsewhere?

Let me stop here and conclude: reliability of our data on Mediterranean plant extinctions is poor. We can hope to improve it in the long run by undertaking a thorough field inventory of the whole flora and by monitoring closely all threatened endemic taxa, but for the time being we are mostly left to guess. Yet, while the contents of extinction lists is bound to undergo considerable change in the future, I would contend that, by and large, their present size will stay. Pending a new qualitative analysis, our quantitative assumptions on plant extinctions in Mediterranean areas are reasonably sound.

6.6 Where do we stand?

Loss of one species in a thousand, as we believe has happened in the Mediterranean area, is not a great deal. Perhaps it is not even significantly more than the natural extinction rate (although one must bear in mind that it happened within little more than a century, which is the average 'age' of Mediterranean species as scientifically described and named taxa). Does this mean that the situation is under control, and that our flora is safe? And, if so, what are the implications of the distinctly higher, yet perhaps not dramatic rates of loss in the other Mediterranean-type floras? The following thoughts are largely hypothetical, yet they may perhaps help to understand the situation we face.

When one compares extinction rates in the four domains with a Mediterranean climate for which such data exist one immediately notes their correlation with the age and duration of large-scale human colonization. The rate is lowest for the Mediterranean proper where agriculture and grazing have now lasted for between 8000 and 6000 years for the most part. It rises as the onset of European settling is delayed, through the 0.3% of the Cape region and the 0.4% of California to the perhaps 1% of the south-western province of Western Australia.

This one might have expected. If one assumes that species elimination peaks in the initial phases of dramatic change induced by human implantation, then much of it may have been over and done with in the Mediterranean by the time the first botanists appeared on the scene and started studying and describing the flora. In the Cape region, the time gap between settlement and exploration was much smaller, in California both may have been more or less concomitant, and in Western Australia the botanist often preceded the settler.

In all these areas the native plant cover disappeared from fertile lands suited for agriculture or grazing. The plants themselves either developed into weeds or succumbed. Since whole vegetation units were eliminated, it is but reasonable to assume that many of their component species shared their fate. Pre-botanical extinction must have been important in the countries bordering the Mediterranean Sea, and may be seen as the price paid for the present relative stability of their flora.

Perhaps one may take the difference between the south-western Australian 1% extinction and the much lower Mediterranean rate as indicative of the degree of unrecorded early extinction in the latter domain—but I rather doubt it. The danger is not yet over for the Australian flora, nor may the present degree of loss be fully understood. Take the genus *Picris* as an example, the Australian representatives of which have just been revised in Berlin. Instead of a single naturalized alien, as was formerly believed, it turned out that Australia houses 10 native species, all but one endemic, plus a couple of alien weeds (Holzapfel and Lack 1993). Of the native ones two or three are no longer extant, having been named and described *post mortem*, and they are all south-western Australian. They are typical examples of species confined to land suited for agriculture or grazing and showing no weed-like colonizing characteristics (Holzapfel 1994). Many more such species doubtless exist or have existed.

Pre-botanical Mediterranean extinction, of which nothing concrete is known, is thus likely to be of a similar order of magnitude as the present and foreseeable extinction caused by man in other Mediterranean-type regions.

6.7 . . . and where to go?

I will limit considerations on the future to the Mediterranean area proper, which is the only region with which I am sufficiently familiar. It is an area in which the flora, no doubt somewhat impoverished but still remarkably rich, has learnt to cope with man and his destructive habits. Present loss rates are tolerably low and can, if one takes care, be further reduced.

This is not the kind of situation in which one asks for priorities among species to be saved: one can and must fight for every single species. The Mediterranean flora had already lost a lot of species, before we began to study it. Losses are now dwindling, and they must be made to stop.

Considerable efforts are certainly needed, but they can mostly be confined to the levels of detailing and monitoring the flora. Traditional human pressures on the natural environment are tolerable and perhaps even necessary, which may pose problems in the not infrequent cases in which land use regresses.

New threats and pressures do unfortunately emerge which the native biota are unprepared to face. Mass tourism affecting the shorelines is probably the worst, but downhill skiing, which is becoming fashionable in some southern European mountainous regions, may come next and cause massive disturbance to hitherto unspoilt natural habitats. Urbanization, industrialization, and consequent pollution are other threats that one must bear in mind. Road, rail, and dam-building may unpredictably affect the single populations of rare endemics unless and until an assessment of their impact is made mandatory before construction may start.

However, all these problems are not of a scale that would make them insoluble. Provided there is sufficient goodwill and concern within the countries concerned—and much progress has recently been made in this respect—plant extinction around the Mediterranean can be successfully checked. Botanical science must play a pivotal role in this process, which requires sound factual knowledge as its very basis.

Summary

Thirty-five species and two subspecies of vascular plants of the Mediterranean area are presumed extinct. This would correspond to an extinction rate of 0.13% of the native Mediterranean flora, which compares to rates of 0.3% for vascular plant species of the Cape floristic province of South Africa, 0.4% for higher plant taxa of California, and 0.66% for those of Western Australia. Percentages of threatened plant taxa are between 25 and 125 times as high as extinction rates. Records of plant extinctions are both incomplete and error-prone, as shown by examples, but even with improving knowledge the rates of species loss are unlikely to change significantly. They are lowest for the Mediterranean area in which human implantation is most ancient and for which large-scale undocumented early extinction is assumed, and highest for the most recently colonized area, south-western Australia, where extinction may now be at its peak. At least for the Mediterranean proper, aiming at the rescue of each and every species in danger is a realistic if ambitious goal.

Acknowledgements

I am grateful to Professor Arne Strid, Dr Gregory Iatrou, and Mr. Sebastian Holzapfel for liberally sharing with me some of their unpublished data, and to Dr

Brigitte Zimmer for her efficient help with the preparation of the tables and manuscript.

References

Anon. (1983). *List of rare, threatened and endemic plants in Europe* (1982 edn). Nature and Environment Series, No. 27. Council of Europe, Strasbourg.

Ayensu, E. S. and DeFilipps, R. A. (1978). *Endangered and threatened plants of the United States.* Smithsonian Institution Press, Washington, DC, and World Wildlife Fund.

Bond, P. and Goldblatt, P. (1984). Plants of the Cape flora. A descriptive catalogue. *J. S. African Bot.* (Suppl.), **13**. Trustees of the National Botanic Gardens of South Africa Kirstenbosch, Claremont.

Conti, F., Manzi, A., and Pedrotti, F. (1992). *Libro rosso delle piante d'Italia.* Associazione Italiana per il World Wildlife Fund, Roma.

Ekim, T., Koyuncu, M., Erik, S., and İlarslan, R. (1989). *Türkiye'nin tehlike altındaki nadir ve endemik bitki türleri IUCN red data book kategorilerine göre hazırlamıştır.* Yayın, No. 18. Türkiye Tabiatını Koruma Derneği, Ankara.

Gamisans, J. and Jeanmonod, D. (1993). Catalogue des plantes vasculaires de la Corse, (2nd edn). In *Compléments au Prodrome de la flore corse,* Série annexe, No. 3, (ed. D. Jeanmonod and H. M. Burdet). Conservatoire et Jardin botaniques de la Ville de Genève, Genève.

Gómez-Campo, C. (ed.) (1987). *Libro rojo de especies amenazadas de España peninsular e islas Baleares.* ICONA, Madrid.

Gómez-Campo, C. (1990). *A germ plasm collection of Crucifers. 1990 list.* Catálogos INIA, No. 22. Instituto Nacional de Investigaciones Agrarias, Madrid.

Greuter, W. (1987). *Onobrychis aliacmonia* (Leguminosae)—the unusual story of a rediscovery. *Pl. Syst. Evol.,* **155**, 215–17.

Greuter, W. (1991). Botanical diversity, endemism, rarity, and extinction in the Mediterranean area: an analysis based on the published volumes of Med-Checklist. *Bot. Chron.,* **10**, 63–79.

Greuter, W., Burdet, H. M. and Long, G. (1984–89). *Med-Checklist. A critical inventory of vascular plants of the circum-Mediterranean countries,* Vols 1–4. Conservatoire et Jardin botaniques de la Ville de Genève and Secrétariat Med-Checklist, Genève and Berlin.

Hadidi, M. N. el, Ghani, M. M. abd el and Fahmy, A. G. (1992). *The plant Red Data Book of Egypt,* 1. *Woody perennials.* Palm Press and Cairo University Herbarium, Cairo.

Hall, A. V. and Veldhuis, H. A. (1985). *South African Red Data Book: plants—Fynbos and karoo biomes.* South African National Scientific Programmes Report, No. 117. Foundation for Research and Development, Pretoria.

Heywood, V. H. (1993). Mediterranean floras and their significance in relation to world biodiversity. In *Connaissance et conservation de la flore des îles de la Méditerranée.* Colloque international, 5–8 octobre 1993, Porticcio—Corse sud. Résumés, abstracts. Parc Naturel Régional de la Corse et Conservatoire Botanique National de Porquerolles, [Ajaccio].

Hickman, J. C. (ed.) (1993). *The Jepson manual. Higher plants of California,* [cancel page for p. 1315]. University of California Press, Berkeley.

Hnatiuk, R. J. (1990). *Census of Australian vascular plants.* Australian Flora and Fauna Series, No. 11. Australian Government Printing Service, Canberra.

Holzapfel, S. (1994). A revision of the genus *Picris* (Asteraceae, Lactuceae) s.l. in Australia. *Willdenowia,* **24**, 97–218.

Holzapfel, S. and Lack, H. W. (1993). New species of *Picris* (Asteraceae, Lactuceae) from Australia. *Willdenowia,* **23**, 181–91.

Lanfranco, E. (1989). The flora. In *Red Data Book for the Maltese Islands* (ed. P. J. Schembri and J. Sultana), pp. 5–70. Environment Division, Ministry of Education, Beltissebh, Malta.

Leigh, J. H. and Briggs, J. D. (ed.) (1992). *Threatened Australian plants. Overview and case studies*. Australian National Parks and Wildlife Service, Canberra.

Lesouef, J.-Y. and Olivier, L. (1989). Bilan de la flore endémique et sub-endémique de France. In *Plantes sauvages menacées de France. Bilan et protection* (ed. M. Chauvet), pp. 119–28. Bureau des Ressources Génétiques and Lavoisier, Paris and Cachan.

Lucas, G. Ll. (ed.) (1980). *First preliminary draft of the list of rare, threatened and endemic plants for the countries of North Africa and the Middle East*. International Union for the Conservation of Nature and Natural Resources, Survival Service Commission, Threatened Plants Committee, Kew.

Phitos, D. (1993). *List of the endemic and rare plants of Greece*. Privately published.

Rothmaler, W. (1944). Floristische Ergebnisse einer Reise nach dem Peloponnes. *Bot. Jahrb. Syst.*, **73**, 418–52.

Smith, J. P. and York, R. (1984). *Inventory of rare and endangered plants of California*. Special Publication, No. 1, ed. 3. California Native Plant Society, Berkeley.

Strid, A. (1986). *Adonis cyllenea* (Ranunculaceae) and *Helichrysum taenari* (Asteraceae) rediscovered in Peloponnisos. *Ann. Mus. Goulandris*, **7**, 221–31.

Strid, A. (1992). *Adonis cyllenea* Boiss. Heldr. & Orph. in Boiss.; *Helichrysum taenari* Rothm, [2 loose sheets without page, date, and place, preprinted from *Greek plant Red Data Book* (ed. D. Phitos)].

Velčev, V., Kožuharov, S., Bondev, I., Kuzmanov, B., Markova, M. and Velev, V. (ed.) (1984). *Červena kniga na NR Bălgarija, Tom* 1. *Rastenija*. Bălgarska Akademija na Naukite, Sofia.

Wraber, T. and Skoberne, P. (1989). Rdeči seznam ogroženih praprotnic in semenk SR Slovenije. *Vrastvo Narave*, **14–15**, 1–429.

7

Recent past and future extinctions in birds

Colin J. Bibby

7.1 Introduction

As a class, birds are second only to terrestrial molluscs in the number of known extinctions in the past few centuries (Jenkins 1992). If we understood extinction processes, we might be able to predict probabilities of future extinctions and target applied research and conservation work. Knowledge of recent past losses of species, either globally or locally, offers some understanding of extinction. Population modelling may give a more rigorous route to insight.

In this chapter, I will examine the extent to which practical guidance for future actions can actually be given on the basis of these approaches.

7.2 Predicting future extinctions

Recent extinctions

About one hundred bird species, or 1% of all birds, are believed to have been lost since 1600 (Temple 1985). Roughly 90% of these were island species, (King 1985; Johnson and Stattersfield 1990). As reviewed by Pimm, Moulton, and Justice in Chapter 5, still more species were lost in the earlier history of human settlement before naturalists ever saw or documented them (see also Diamond 1982; Olson 1989). Milberg and Tyrberg (1993) describe over 200 extinct island species known from sub-fossil remains in spite of the large numbers of islands yet to be explored.

Factors salient in the extinction record on islands include human predation (especially of large flightless species), introduced predators (cats, rats, dogs, mongooses, monkeys, etc), habitat loss due to direct human impact or introduced grazers and browsers (goats, pigs, cattle, and rabbits), and introduced diseases. Species which evolved in the absence of natural predators on remote oceanic islands have often become flightless (Diamond 1981), apparently tame, and otherwise lacking in predator avoidance. Such species may also be vulnerable to introduced disease as suggested on Hawai'i (Ralph and van Riper 1985). Often, island species are further vulnerable by virtue of small populations and ranges. The relatively uniform habitats on islands lower the chances of population refugia surviving and recolonizing (Frankel and Soulé 1981). While many oceanic island species are already gone, a highly vulnerable class survives on those islands which have not yet been colonized by alien predators (Atkinson 1985; Moors *et al.* 1992).

Local losses from habitat islands

Comparable losses have yet to occur amongst continental species although they are widely predicted to be imminent (e.g., Ehrlich 1986; Myers 1989). Many species have small total ranges (about 27% of all birds have total ranges below 50 000 km^2, ICPB 1992). Large areas of natural habitat, most strikingly tropical forests, have been cleared in many countries. As a result, an increasing number of continental species with naturally small ranges are now isolated in shrinking habitat islands (Terborgh 1974). Factors leading to loss of local populations may not be the same as those causing global extinctions (Soulé 1983). On the other hand, many currently threatened birds are confined to dwindling numbers of local populations (e.g., Collar *et al.* 1992). In such cases, global extinction would merely be the last of a series of losses of local populations.

Island biogeographic theory has been used to infer that fragmented habitat patches will lose species. Direct observation of this phenomenon has been made in Brazilian coastal forests (Terborgh and Winter 1980), on the artificial island of Barro Colorado (Karr 1982), and in experimentally created fragments in Brazil (Bierregaard *et al.* 1992).

Salient predictors of extinction in small habitat patches include initially small population size, vulnerability to predators, membership of specialist guilds (such as large predators, army-ant followers or other mixed species flocks), dependence on variably available diets (such as fruits or nectar), larger members of guilds, and local catastrophes (such as hurricanes). Vulnerability to predators is often exacerbated by the disappearance of the largest species, which in general do not prey on birds but often kill smaller predators. In their absence, smaller predators with differing diets, including birds, may become more abundant (Terborgh 1988).

Small population models

Empirical observations are of limited use in making predictions because so many species share some of the factors known to have been associated with global extinction or local losses (Simberloff 1986). It is likely that factors acting in combination will be more lethal, but how do we pick the species which, in the absence of intervention, would actually head the list of extinctions in this century?

Population viability analysis (PVA) is a process which might help (Green and Hirons 1991). Indeed, new proposals for categorizing threatened species are based on categories of probability of extinction over given time periods (Mace, Chapter 13; see also Mace and Lande 1991, Mace *et al.* 1993). The interpretation of model outputs is open to various criticisms (Boyce 1992). The process requires ecological and life-history data, a reasonable model, and insight into the factors which might change in the future, either naturally or within a management regime. For such models to be of practical use, we usually require data about reproductive rates and their variances, survival rates and their variances, density dependence relationships for survival and reproduction, frequency of catastrophic events and their impact, carrying capacities of habitats and their variances, and dispersal rates between sub-populations.

There is a serious practical difficulty in obtaining many of these measures, especially if the species is endangered and urgency precludes long-term study. Variance estimates can only be acquired over a run of time which may simply not be available. Catastrophes, which may be devastatingly quick in their effect, can easily be imagined. But how do you estimate the probability of a cat being landed on a particular small island in the next five years?

This is not to suggest that PVAs are worthless. The PVA process entails bringing experts and managers together to agree on the meaning of available information and on likely management options and to explore possible consequences. The prospect of a continuous process of modelling, management, and observation supports the ideal notion of management regimes being adaptive to changing knowledge and circumstances. The majority of conservation decisions are still taken in the absence of formal population viability analysis and this is likely to continue to be the case. A move to greater formality would be a great aid in focusing scientific attention on the kinds of data most urgently needed to improve the chances of success of a management plan.

7.3 Red Data Books

The basis of Red Lists for birds

As discussed in more detail by Mace in Chapter 13, one approach to predicting which species are at risk of extinction has been the IUCN-promoted Red Lists. These have reached their greatest elaboration for birds, which are the only class amongst which all species have been reviewed and classified as 'threatened', 'near threatened' or 'safe' (Collar and Andrew 1988). In the case of Africa (Collar and Stuart 1985) and the Americas (Collar et al. 1992), extensive literature reviews have been published. We might ask how accurate these lists are in predicting future candidates for imminent extinction. A more practical question might be to ask what conservation biologists should next be doing to help.

Threatened species of the Americas, including the near Pacific islands and the Caribbean, have been listed five times (Anon. 1964; Vincent 1966–71; King 1981; Collar and Andrew 1988; Collar et al. 1992). Over the last 30 years, the list has expanded fivefold (Table 7.1) with most of the growth being in continental South America.

By 1988 the list for the Americas had grown to 360 species. This list was reviewed in greater depth leading to the 1992 publication. These two most recent reviews differ by 141 species (Table 7.2). Some (24) of the changes were due to taxonomic alterations or discoveries. In general, the consultative process producing the 1988 list tended more towards precautionary listing of species than

Table 7.1 Numbers of threatened bird species listed for the American region at various dates. The area covered has been standardized to that in Collar *et al.* (1992). (Sources are given in the text.)

	1964	1971	1979	1988	1992
North America	10	6	7	13	12
Pacific islands	9	10	4	19	15
Caribbean	18	13	16	31	37
Latin America	31	27	62	297	263
Totals	68	56	89	360	327

Table 7.2 Changes of classification of threatened species in the Americas between 1988 and 1992 and their causes

	No. of species
Listed in 1988	360
Listed in 1988 but not 1992	−87
Listed in 1992 but not 1988	+54
Listed in 1992	327

Deletions from the 1988 list
6 species no longer regarded as valid
76 species regarded from new knowledge as 'near threatened'
5 species regarded from new knowledge as 'safe'

Additions in the 1992 list
6 species newly discovered since 1988
20 species newly erected by taxonomic split or clarification
11 species missed from earlier review
18 species reassessed as a result of better knowledge (of which 14 had been listed as 'near threatened')

to omission of candidates found, on fuller study, to be valid (Collar, pers. comm.). Of 29 species entering the threatened list for the first time in 1992, 14 had previously been indicated as near threatened and only 15 had been inappropriately, as revealed by fuller review, omitted or overlooked.

The general similarity of these two most recent listings suggests a convergence of opinion on which the threatened species are. The very process of reviewing and listing species with documentary evidence can help to promote and target further study. Ideally, knowledge gaps would be indicated and people would be stimulated to fill the most urgent in a continuous process of discovery and review. There is some evidence that this has actually happened.

Table 7.3   Quality of knowledge of birds of the Americas listed as threatened in 1992

		No. of species
Trend in the last 50 years		
(0)	Unknown	121
(1)	Inferred downwards because of habitat loss	133
(2)	Inferred downwards from decrease in numbers of records	50
(3)	Quantified in the last 20 years	23
Numbers		
(0)	Totally unknown	76
(1)	Inferred rare from few records or small range	173
(2)	Some counts or densities	41
(3)	Total population estimate	37
Range		
(0)	Virtually unknown no pattern, few recent records	37
(1)	Some pattern with gaps—few known current sites	121
(2)	Fairly well known in several to many current localities, but new discoveries likely	110
(3)	Formerly bounded, potentially measurable, new localities unlikely	59

Erroneous judgement is clearly less likely if population parameters of species are adequately documented (King 1987). The degree of knowledge of range, population, and trend of threatened species listed in 1992 is indicated in Table 7.3. In spite of a huge upsurge of ornithological interest in Latin America, the state of knowledge of many birds remains poor, as shown by Green and Hirons (1991).

Of the three parameters considered, range tends to be best known, but many species have been recorded from rather few localities and can only be presumed to be likely to occur in nearby areas of similar habitat (if there are any surviving). Less than a quarter of threatened species have been subject to any formal counting. The majority are inferred to be rare because they have infrequently been seen within what are often known or inferred to be very limited ranges. Not surprisingly, formal estimates of population trend are still less common because even fewer species were counted 20 years ago. The most frequent pattern of trend is that numbers are inferred to be declining because of habitat loss which has often been very extensive within known ranges.

There is some relationship between quality of knowledge and category of threat (Table 7.4). A slightly higher proportion of endangered species are relatively well known, but this is primarily because of the inclusion of some very thoroughly studied North American species. On the other hand, 10 endangered species are virtually unknown. Species categorized as rare are often better known but the causation may be that they have to be reasonably well known before they can be classified as rare rather than insufficiently known.

Table 7.4 Distribution of quality of knowledge of threatened species in the Americas (1992) in relation to threat class. Degree of knowledge for each species is the sum of three scores (0–2) shown in Table 7.3

Class of threat	Quality of knowledge				Mean	n
	0	1–3	4–6	7–9		
Endangered	10	32	36	18	4.1	96
Vulnerable	4	24	21	3	3.4	52
Vulnerable/rare	1	35	37	6	3.8	79
Indeterminate	10	26	12	2	2.6	50
Rare	1	4	8	6	5.2	19
Insufficiently known	2	20	8	1	2.6	31
Total	28	141	122	36		327

Table 7.5 Frequency of classes of recommendations made for the conservation of 327 threatened bird species in the Americas

Action	No. of species
Secure sites	241
Locate new sites	214
Estimate populations in sites	197
Study ecology	164
Manage sites	91
Control taking	49
Educate people	44
Captive management	23
Taxonomic study	8
Other	14

Recommendations from Red Data Books

The recommended actions from *Threatened birds of the Americas* (Collar *et al.* 1992) are summarized in Table 7.5. Within the divisions made, there are, in total, 1045 recommendations for 327 species. A high proportion (71%) concern a set of four site-specific actions, (find new sites, protect sites, manage them and enumerate populations within them). For only five species do the recommendations include no site-based actions. Species were classified according to degree of threat and reasons for the assessment (see Collar *et al.* 1992). Recommended actions tend to be rather similarly distributed according to threat class ($\chi^2 = 77.0$; d.f. $= 55$).

The cause of such results is that the most frequent pattern for threatened birds in Latin America combines limited distribution, dramatic habitat loss, and limited protection of representative areas of the habitat. Minimum conservation measures in such circumstances entail finding populations if the locations of few (or none) are known, followed by protection of viable populations in representative sites. Since numbers are so rarely known, there is a major challenge to survey the threatened species in protected areas as a first step towards establishing whether populations are likely to be viable. Ecological study might assist the management process in protected areas by diagnosing causes of decline and suggesting remedies.

The tropical pattern of risk of extinction differs from that in northern temperate latitudes where, both absolutely and relative to the smaller avifaunas, fewer species are globally at risk. The few that are globally threatened have often been recorded as declining over long periods of time.

7.4 Biodiversity approaches

The fact that so many threatened species have very limited ranges leads to the obvious priority for identifying and protecting the uniquely important sites where they occur (Terborgh and Winter 1982). In a global study (ICBP 1992), we have shown that 27% of all birds on Earth (2609 species) have breeding ranges of less than 50 000 km^2. There is a high degree of overlap among these ranges which have been classified into 221 distinct areas of endemism, each with at least two unique restricted-range species. By this definition, endemic bird areas (EBAs) embrace about 5% of the Earth's land surface and uniquely accommodate 2484 (95%) restricted-range birds.

In the continental Americas with their nearby islands (the same area as covered by Collar et al. 1992), 81 EBAs occupy just over 3 million km^2 embracing the full range of 999 bird species. Of these restricted-range species, 253 are listed as threatened. This amounts to 77% of the threatened birds of the region.

In addition, a further 52 threatened species with ranges greater than 50 000 km^2 also occur in EBAs (Table 7.6). Thus the future of 90% of the threatened birds of the Americas could largely be determined by the sustained existence of sufficient natural habitat in these EBAs. Of the species not embraced by EBAs, 17 have limited ranges not co-incident with other such species and 4 are sea birds in need of protection at their limited breeding sites. Thus, these 21 species could also largely be safeguarded by appropriate actions in very limited areas. Of just 16 remaining widespread and threatened species, 7 occur in the United States.

The potential efficiency of managing threatened species by conservation policy in EBAs can be illustrated by the areas involved. If attention was given to threatened species, all 253 are confined to 2.2 million km^2 but 209 to only 1 million km^2 (Fig. 7.1).

The extent to which individual threatened species with restricted ranges are found in protected areas is one measure of the degree to which their poor conservation status has been recognized and acted upon. On the other hand, if

Table 7.6 Occurrence of threatened birds in the Americas in relation to range and endemic bird areas (EBAs)

Distribution	No.
Restricted-range and solely in EBAs	237
Restricted-range occurring singly	18
Range > 50 000 km² but occurring in EBAs	52
Others (4 sea birds, 7 US, and 9 South American species)	20
Total	327

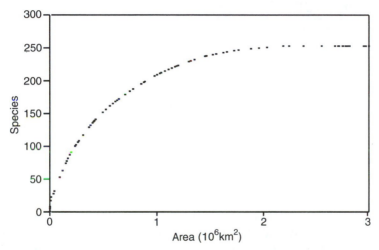

Fig. 7.1 Numbers of threatened species in relation to land area in 81 endemic bird areas (EBAs) in the Americas. Species number and land area are summed cumulatively across EBAs ranked in order of declining numbers of species per unit area.

such species do not occur in protected areas, it is very likely that other elements of biodiversity, both species and communities, will also be unprotected. Knowledge of the distribution of threatened species can thus be used to assess the degree to which they are threatened and also the adequacy of existing networks of protected areas.

An example is shown in Fig. 7.2. The Albertine Rift mountains in Central Africa have 41 endemic birds and can be divided into 12 separate areas of surviving forest, of which 7 enjoy some formal protection. Not surprisingly, the 15 threatened species are still less widespread than the other local endemics, occurring, on average, in only two areas each. At the moment, 9 threatened species do not occur in a protected area at all. Had the protected areas been chosen differently, a set of five would have been sufficient to secure at least one

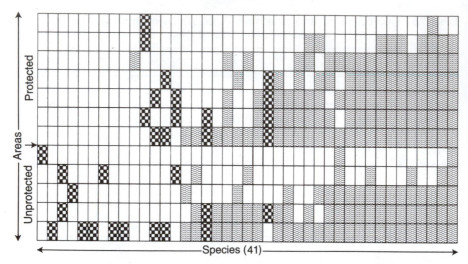

Fig. 7.2 Occurrence of 41 species endemic to the Albertine Rift mountains (Central Africa) in relation to 12 subdivisions of the region. Threatened species are hatched in bold.

place of occurrence for all the threatened species. Four of these five are the unique locations for one threatened species each and are not currently protected.

7.5 Discussion

Evidence from previous extinctions, losses from habitat fragments, and population modelling all point to small populations and losses of habitat as recurring themes on the road to extinction. Birds with these features can be identified fairly readily and include about 10% of all the world's species. The impact of stochastic variation on extinction risk is clear from modelling and from observed losses from habitat fragments. It is rather unclear how vulnerable most particular species or habitats are to stochastic variation beyond the observation that fruit and nectar feeders are vulnerable. Both modelling and observation point to the devastating potential of catastrophic events. With the exception of hurricanes, it is difficult to see how to anticipate the identity of most possible catastrophes, let alone their probability. This is most evident in the case of islands. If the predator-vulnerable species still survive, there is a perpetual risk of accidental introductions. The impact of an alien snake on Guam (Savidge 1987) shows how steep the slope to extinction can be. In less than 40 years, the island lost 7 of its 25 species, 4 survive in small numbers, and 7 more have declined. The Guam flycatcher *Myiagra freycineti* declined from 450 individuals in 1981 to extinction by 1984 (Engbring and Pratt 1985). This is the fastest rate of decline of any threatened species for which such an estimate could be made (Green and Hirons 1991). Only perpetual precaution and vigilance can lower the chances of such disaster and warn of the need for urgent action should it occur. Clearly, the warning time can be very brief.

The coincidence in range of many rare species argues strongly for the identification, designation, and adequate protection of representative areas. Systematic review of the threatened birds of the Americas (Collar *et al.* 1992) provides a very clear indication of priority species and places to start such a search. A further analysis (Wege and Long 1994) portrays these priorities with a clearer geographic rather than species-based emphasis.

Evidence reviewed by Thirgood and Heath (1994) points to the strong likelihood that centres of endemism for birds will be centres of endemism for other taxa at a broad scale. There is a striking, but not complete, similarity between the formally identified endemic bird areas and centres of plant diversity which have been identified semi-formally (Anon. 1992; Davis and Heywood in prep.). While we might hope for a great increase in knowledge of bird species in the next few years, this is optimistic for the majority of the world's other species which may not be identified for a few hundred years. The use of indicator species from other taxa would be a prudent approach to identifying gaps in protected area coverage missed by ornithologists.

Although the majority of species which have so far been driven to extinction would always have been relatively rare by virtue of limited range and particularly vulnerable because threats could have operated throughout their small ranges, some spectacularly abundant species (e.g., passenger pigeon *Ectopistes migratorius* and Carolina parakeet *Conuropsis carolinensis*) have also been exterminated. Concentration on species at risk of extinction also ignores the progressive erosion of biodiversity represented by reduction of range and numbers of many species. Imboden (1987) and Diamond (1987) have therefore proposed the green listing of those species which are ubiquitous and tolerant of habitat modification.

Such an approach has been taken in Europe where nearly half of all species have an unfavourable conservation status, although rather few of them are currently at risk of global extinction (Tucker *et al.* 1994). It would be wise to review the conservation status of all the world's birds and draw attention to widespread declines as well as risk of extinction.

Summary

Most recent extinctions of birds have been caused by habitat loss or by human or introduced predators and have been on islands. Local losses of species in habitat patches are particularly prevalent amongst various specialist feeders and species occurring in small numbers. Future candidates for global extinction are hard to pick from lists of species with indicators of susceptibility. Population modelling should help, but data are generally lacking.

Review of threatened birds in the Americas shows that declines and rarity are often inferred from habitat loss and infrequent records, in the absence of quantitative data. The most threatened species often occur in very few places, where their future is likely to be determined. Safeguarding protected areas within centres of endemism offers a pragmatic response for a high proportion of globally threatened birds and probably other taxa as well.

Acknowledgements

I gratefully appreciate helpful discussion of ideas and comment on a draft from Michael Crosby, Adrian Long, Alison Stattersfield, David Wege, and especially Nigel Collar.

References

Anon. (1964). *List of rare birds, including those thought to be so but of which detailed information is still lacking*. International Union for Conservation of Nature and Natural Resources, Morges, Switzerland.

Anon. (1992). Centres of species diversity. In *Global biodiversity: status of the earth's living resources*, (ed. B. Groombridge), pp. 154–91. Chapman & Hall, London.

Atkinson, I. A. E. (1985). The spread of commensal species of *Rattus* to oceanic islands and their effects on island avifaunas. In *Conservation of island birds*, Technical Publication, No. 3, (ed. P. J. Moors), pp. 35–81. International Council for Bird Preservation, Cambridge.

Bierregaard, R. O., Lovejoy, T. E., Kapos, V., Augusto dos Santos, A., and Hutchings, R. W. (1992). The biological dynamics of tropical rainforest fragments. *BioScience*, **42**, 859–66.

Boyce, M. S. (1992). Population viability analysis. *Ann. Rev. Ecol. Syst.*, **23**, 481–506.

Collar, N. J. and Andrew, A. (1988). *Birds to watch: The ICBP world check-list of threatened birds*, Technical Publication, No. 8. International Council for Bird Preservation, Cambridge.

Collar, N. J. and Stuart, S. N. (1985). *Threatened birds of Africa and related islands: The ICBP/IUCN Red Data book*. International Council for Bird Preservation and International Union for the Conservation of Nature and Natural Resources, Cambridge.

Collar, N. J. *et al.* (1992). *Threatened birds of the Americas: The ICBP/IUCN Red Data Book*. International Council for Bird Preservation, Cambridge.

Davis, S. D. and Heywood, V. H. (ed.). *Centres of plant diversity: a guide and strategy for their conservation*. Gland: International Union for Conservation of Nature and Natural Resources, in prep.

Diamond, J. M. (1981). Flightlessness and fear of flying in island species. *Nature*, **293**, 507–8.

Diamond, J. M. (1982). Man the exterminator. *Nature*, **298**, 787–9.

Diamond, J. M. (1987). Extant unless proven extinct? Or, extinct unless proven extant. *Conserv. Biol.*, **1**, 77–9.

Ehrlich, P. H. (1986). Extinction: what is happening now and what needs to be done. In *Dynamics of extinction*, (ed. D. K. Elliott), pp. 157–64. Wiley, New York.

Engbring, J. and Pratt, H. D. (1985). Endangered birds in Micronesia: their history, status and future prospects. *Bird Conserv.*, **2**, 71–105.

Frankel, O. H. and Soulé, M. E. (1981). *Conservation and evolution*. Cambridge University Press.

Green, R. E. and Hirons, G. J. M. (1991). The relevance of population studies to the conservation of threatened birds. In *Bird population studies*, (ed. C. M. Perrins, J.-D. Lebreton, and G. J. M. Hirons), pp. 594–633. Oxford University Press.

ICBP (International Council for Bird Preservation) (1992). *Putting biodiversity on the map: priority areas for global conservation*. International Council for Bird Preservation, Cambridge.

Imboden, C. (1987). Green lists instead of Red Books? *World Birdwatch*, **9**, 2.

Jenkins, M. (1992). Species extinction. In *Global biodiversity: status of the earth's living resources*, (ed. B. Groombridge), pp. 192–205. London: Chapman & Hall.

Johnson, T. H. and Stattersfield, A. J. (1990). A global review of island endemic birds. *Ibis*, **132**, 167–80.

Karr, J. (1982). Avian extinctions on Barro Colorado Island, Panama: a reassessment. *Amer. Nat.*, **119**, 228–39.

King, F. W. (1987). Thirteen milestones on the road to extinction. In *The road to extinction: problems of categorizing the status of taxa threatened with extinction* (ed. R. Fitter and M. Fitter), pp. 7–18. International Union for Conservation of Nature and Natural Resources, Gland, Switzerland and Cambridge.

King, W. B. (1981). *Endangered birds of the world: the ICBP bird Red Data Book*. Smithsonian Institution Press, Washington, DC, in cooperation with International Council for Bird Preservation, Cambridge.

King, W. B. (1985). Island birds: will the future repeat the past? In *Conservation of island birds*, Technical Publication No. 3, (ed. P. J. Moors), pp. 3–15. International Council for Bird Preservation, Cambridge.

Mace, G. M. and Lande, R. (1991). Assessing extinction threats: toward a reevaluation of IUCN threatened species categories. *Conserv. Biol.*, **5**, 148–57.

Mace, G. M. *et al.* (1992). The development of new criteria for listing species on the IUCN Red List. *Species*, **19**, 16–22.

Milberg, P. and Tyrberg, T. (1993). Naive birds and noble savages—a review of man-caused prehistoric extinctions of island birds. *Ecography*, **16**, 229–50.

Moors, P. J., Atkinson, I. A. E., and Sherley, G. H. (1992). Reducing the rat threat to islands. *Bird Conserv. Internat.*, **2**, 93–114.

Myers, N. (1989). A major extinction spasm: predictable and inevitable? In *Conservation for the twenty-first century*, (ed. D. Western and M. Pearl), pp. 42–9. Oxford University Press.

Olson, S. L. (1989). Extinction on islands: man as a catastrophe. In *Conservation for the twenty-first century*, (ed. D. Western and M. Pearl), pp. 50–3. Oxford University Press.

Ralph, C. J. and van Riper, C. (1985). Historical and current factors affecting Hawaiian native birds. *Bird Conserv.*, **2**, 7–42.

Savidge, J. A. (1987). Extinction of an island forest avifauna by an introduced snake. *Ecology*, **68**, 660–8.

Simberloff, D. (1986). The proximate causes of extinction. In *Patterns and processes in the history of life*, Life Sciences Research Report, No. 36, (ed. D. M. Raup and D. Jablonski), pp. 259–76. Springer, Berlin.

Soulé, M. E. (1983). What do we really know about extinction? In *Genetics and conservation*, (ed. C. M. Schonewald-Cox, S. M. Chambers, B. MacBryde, and W. L. Thomas), pp. 111–24. Benjamin/Cummings, Menlo Park, CA.

Temple, S. A. (1985). The problem of avian extinctions. In *Current ornithology*, (ed. R. F. Johnston), Vol. 3, pp. 453–85. Plenum, New York.

Terborgh, J. (1974). Preservation of natural diversity: the problem of extinction prone species. *BioScience*, **24**, 715–22.

Terborgh, J. (1988). The big things that run the world—a sequel to E. O. Wilson. *Conserv. Biol.*, **4**, 402–3.

Terborgh, J. and Winter, B. (1980). Some causes of extinction. In *Conservation biology: an evolutionary-ecological perspective*, (ed. M. E. Soulé and B. A. Wilcox), pp. 119–33. Sinauer, Sunderland, MA.

Terborgh, J. and Winter, B. (1982). Evolutionary circumstances of species with small ranges. In *Biological diversification in the tropics*, (ed. G. T. Prance), pp. 587–600. Columbia University Press, New York.

Thirgood, S. J. and Heath, M. F. (1994). Global patterns of endemism and the conservation of biodiversity. In *Systematics and conservation evaluation*, (ed. C. J. Humphries and R. I. Vane-Wright), pp. 207–27. Oxford University Press.

Tucker, G. M., Heath, M. F., Tomialojc, L., and Grimmett, R. F. A. (1994). *Birds in Europe: their conservation status*. BirdLife International, Cambridge.

Vincent, J. (1966–71). *Red Data Book*, 2. *Aves*. International Union for Conservation of Nature and Natural Resources, Morges, Switzerland.

Wege, D. C. and Long, A. J. (1994). *Priority Areas for threatened birds in the Neotropics*. BirdLife International, Cambridge.

8

Rates and patterns of extinction among British invertebrates

J. A. Thomas and M. G. Morris

8.1 Introduction

With about 22 000 species of insect and several thousand other invertebrates, Britain has a comparatively small invertebrate fauna and few endemics to compensate for the paucity (Kerrich *et al.* 1978). However, Britain is better known than any other comparable area in the world (Bratton 1991; Collins and Thomas 1991), and is probably the only country for which the rates and causes of species loss over the past 100–300 years can be reviewed. Our conclusions, although tentative, apply to much of Europe (Heath 1981; Erhardt and Thomas 1991; Speight 1989; Thomas 1991; Beek *et al.* 1992) and perhaps elsewhere. For whereas European biotopes were once considered atypical because of management and alteration by man (Duffey and Watt 1971; Rackham 1986; Spellerberg *et al.* 1991; Warren and Key 1991; Thomas 1993), evidence is accumulating that many biotopes in other continents have been similarly shaped by man during recent millennia (e.g., Hammond and McCorkle 1984; Pimm, Moulton, and Justice, Chapter 5). In studying how the invertebrates of low-input low-output agricultural and forestry systems have reacted to the agrichemical and silvicultural revolutions of the 20th century, we may illuminate processes to be repeated over much of the world as human populations grow and societies 'develop'.

In Britain, concern about declining invertebrates began in the 19th century with entomologists lamenting local declines of butterflies. Popular campaigns are very recent, yet so influential that three proposed motorway routes have been altered principally to avoid destroying butterfly colonies (e.g., Munguira and Thomas 1992). But the greater the prize, the more apt are campaigners to make unsubstantiated claims about extinction rates. We examine the evidence for decline in historical times, and describe how modern recording schemes should provide a more reliable database for future assessments. We also seek patterns in the declines of species, and to understand the controlling mechanisms, believing that this will provide a more reliable basis than extrapolation for predicting which species may decline in future.

8.2   Estimating extinction rates of British invertebrate species

Historical records

Species lists and status accounts for British invertebrates over the past 200–300 years are unmatched elsewhere, but are far less complete than most biologists suppose. For the best studied group, the butterflies, a national list of 20 species was compiled in 1666, and nearly half the species had been discovered by 1710 (Ford 1945): but the list was completed only in the late 19th century (Fig. 8.1.) (a 20th-century addition came from recognition that one species was two). Discoveries of Odonata closely followed those of butterflies while other taxa remained comparatively unknown: 18% of Orthoptera and about 25% of freshwater and terrestrial mollusc species now recognized as native (excluding casuals and recently established aliens) were discovered here during the present century (Marshall and Haes 1988; South 1974). In most other groups, such as woodlice (Fig. 8.1): 'we are still adding to the British list rather than detecting species that are becoming extinct' (P. T. Harding, pers comm.).

Useful accounts of distribution and status generally lagged behind species lists. Freshwater and terrestrial molluscs are an exception: a pioneering census of known species was published by the Conchological Society in 1885, with six further editions up to 1951, containing distributions shown by vice-counties (vice-counties are subdivisions of the old British counties, often used in compiling lists of species) (South 1974). More appropriate scales for assessing change were used by Kerney (1976). With butterflies, the first thorough compilation of distributions and status was made just 80 years ago, and although the information encompassed the entire 19th century, it was restricted to local species (Tutt 1905–14). Comparable data for common species were obtained only after

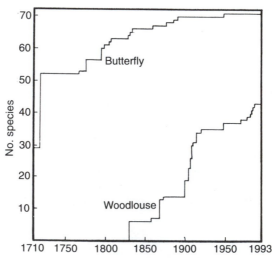

Fig. 8.1   Dates at which species of butterfly and woodlouse were discovered in Britain.

the Biological Records Centre (BRC) launched a national mapping scheme for butterflies in 1967 (Heath *et al.* 1984).

Comparatively complete historical records also exist for Odonata (Moore 1976) and water beetles (Foster 1991). Information on other taxa is extremely patchy, although there are many anecdotal and other data in books, journals, and museums, from which inventories for some counties and many localities can be compiled. Much dates from between the mid-19th century and the First World War, when insect collecting was popular.

Modern databases

The modern era of invertebrate recording began around 1960 (Perring 1992) with the establishment of many national and local mapping schemes, monitoring schemes, and site and species surveys. At a national scale, the BRC co-ordinated the data collecting of several zoological societies (Harding 1992). Forty-seven national recording schemes currently exist, encompassing 10 000 species in 22 orders of invertebrate (Harding and Sheail 1992). The database contained 1.25 million records by 1989, and has resulted in 18 national atlases of invertebrate taxa, with 6 more scheduled for 1994/1995.

Local recording centres have been equally prolific: about 70 existed by 1987, mainly based at county or city museums (Harding and Sheail 1992). Although interfacing between national and local centres remains a problem, these local databases have enabled (usually county) maps of changing distributions to be produced at the scale of tetrads or 1 km squares. Already, at least 32 local atlases of butterflies have been published since 1983. As with several national atlases, most authors also examined a variety of obscure historical sources to make authoritative statements on species changes over the last century or more.

In addition, intensive site surveys were organized by the Nature Conservancy Council (NCC) and resulted in a National Site Index for Notable Invertebrates; the National Trust runs a parallel scheme (Alexander 1992). National surveys of endangered species have been organized mainly by a Joint Committee for the Conservation of British Insects, and involve using a standardized method (Thomas 1983*a*) to estimate the boundaries and size of every surviving population.

Finally, three monitoring schemes have been established, sampling aphids, macro-moths, and butterflies. All record population changes of species from one generation to the next on fixed sites throughout the British Isles. The results enable conservationists to detect long-term declines or increases in population size before these become apparent through mapping schemes as extinctions or spreads; they also enable scientists to study patterns in abundance and change. Rothamsted has monitored aphids and moths since the 1960s (Woiwod and Harrington 1994), and Monks Wood, butterflies since 1976 (Pollard and Yates 1993).

Thus, modern monitoring and recording schemes are creating a vast database from which future changes in status can be gauged. But substantial gaps remain: the parasitic Hymenoptera (about a quarter of the British insect fauna) have negligible cover, as do many Diptera and some Coleoptera. For these and other obscure groups, the current dearth of young professional taxonomists means that the past 25 years may yet be seen as a brief golden age in British recording rather than the start of an era of scientific recording. Funding for most schemes is also uncertain.

Assessing changes in the status of British invertebrates

Variation in recording effort between taxa and over time has been so great that formidable difficulties exist in interpretation. We would expect a greater proportion of British butterfly species to be listed as extinct (Table 8.1) simply because rarities, which disappeared > 100 years ago, were recorded in this group but not in others. Substantial bias still exists if the same timescale is used, due to differences in the intensity and accuracy of past recording; moreover, some Victorian records were fraudulent (Allan 1943). Even within popular groups, rare, spectacular, and accessible species, regions and biotopes are greatly over-represented in the records, which frequently leads to mistaken conclusions by non-specialists. And too often, changes in range are equated with changes in status, again with misleading results.

Nevertheless, by the late 1970s there was consensus that British insect databases were adequate for 95 specialists to review their specialist groups (apart from Microlepidoptera, Parasitica, and some Diptera) using the International Union for the Conservation of Nature and National Resources (IUCN) classifications of Extinct, Endangered, etc. (Mace, Chapter 13). Their conclusions were submitted to a Red Data Book (RDB) selection committee, which aimed for consistency between taxa. The known habitats and perceived threats were also described, and published with assessments, such as the *British Red Data Books*, 2. *Insects* (Shirt 1987). Other invertebrates have been similarly reviewed (Bratton 1991), covering nine major taxonomic groups but excluding Protozoa. Subsequent reviews have been made of Ephemoroptera and Plecoptera (Bratton 1990), spiders (Merrett 1990), most Coleoptera (Hyman and Parsons 1992), and pyralid moths (Parsons 1993).

The percentage of each group of invertebrates considered to be 'extinct', 'endangered' or 'vulnerable' is summarized in Table 8.1. For several taxa, the proportion of threatened species is similar to that in vascular plants (about 7%; Harding and Sheail 1992). However, column 3 of Table 8.1 undoubtedly underestimates extinctions in taxa that were poorly recorded before the 20th century (i.e., all except butterflies and dragonflies). Although several butterfly species inhabit one of the most rapidly disappearing types of habitat within Britain, it is unlikely that they have declined more over the past 250 years than the aculeate Hymenoptera, which are even more characteristic of the same habitat-type (sections 8.3, and 8.4) (Falk 1991). The percentages of species in all RDB categories may more accurately assess the comparative vulnerability of different taxa (Table 8.1, column 5).

We also compared the extinctions of vertebrates, butterflies, and dragonflies in the county of Suffolk, which has an exceptionally thorough history of wildlife

Table 8.1   Percentage of invertebrate species classified in British Red Data Books and other reviews

Order	No. of British species	Presumed 'Extinct' (%)	'Vulnerable' or 'Endangered' (%)	In Red Data Books (%)
Insects				
Odonata (dragonflies)	41	7	7	22
Orthoptera (grasshoppers, crickets)	30	0	17	20
Heteroptera (bugs)	540	1	4	15
Trichoptera (caddis flies)	19	2	6	17
Lepidoptera (butterflies)	59	8	5	20
(macro-moths)	c.900	2	3	11
Coleoptera (beetles)	c.3900	2	6	14
Aculeate Hymenoptera (ants, bees wasps)	580	4	7	28
Diptera (flies)	c.6000	0.05	8	14
Others				
Mollusca (slugs, snails)	c.200	0	9	17
Annelida (leeches)	16	0	0	19
Myriapoda (centipedes, millipedes)	c.88	0	0	6
Crustacea (woodlice, amphipods)	64	0	0	9
Arachnida (spiders, pseudoscorpions)	c.665	0	8	13

recording (Table 8.2). This suggests that invertebrates may have experienced more species extinctions than other popular groups, with butterflies being more vulnerable than dragonflies. In corroboration, Thomas (1991) described the extinction of many butterfly species in semi-natural biotopes where the flora is apparently unchanged. In the Netherlands, where agriculture is equally intensive, 21% of butterfly species have become extinct in historical times and 46% are in long-term decline; similar declines have occurred among Orthoptera and mammals, but those of birds, lower plants, and some groups of vascular plants are lower (van Swaay 1990).

We conclude that the extinction rate of butterfly species in Britain has probably exceeded those of vascular plants or vertebrates in historical times, and infer that some less popular orders, like aculeate Hymenoptera (Falk 1991) and Orthoptera (Marshall and Haes 1988), may have experienced similarly acute losses. Certainly, several reviewers found that the extinctions and declines in their specialist group were considerably greater than had previously been thought (e.g., Falk 1991). On the other hand, some media campaigns have grossly overstated declines: one example is an advertising campaign run by the Butterfly

Table 8.2   Extinction of species in Suffolk. (From Thomas 1991 and
Mendel 1992.)

Species	No. of species (c.1850)	No. of species (1980s)	Change (%)
Butterflies	50	29	−42
Dragonflies	26	22	−15
Vascular plants	1418	1343	−5
Amphibia and reptiles	9	8	−12
Mammals	35	34	−3
Birds	114	130	+14

Conservation Society, which we believe was wholly unjustifiable on scientific grounds (although successful in that membership rose from 3000 to > 10 000 in three years).

During the same recording periods, a few species in most taxa have increased in Britain. About 9% of butterfly species increased this century compared with 74% that declined (Heath *et al.* 1984). The ratio is similar in other groups (Falk 1991).

Predicting future extinctions from current trends

'Vulnerable' or 'endangered' status (Table 8.1) implies extinction for a species in Britain in the near or foreseeable future if factors causing its decline continue. More precise predictions of the date of extinction have been made for three species of butterfly by extrapolating the steep and steady extinction rates of previous decades (Thomas 1983*b*; Thomas *et al.* 1986; Warren *et al.* 1984). In two cases this was a happy failure. Predictions for *Lysandra bellargus* and *Hesperia comma* were made at their nadir, and today nearly twice as many populations exist as 10 years ago (Thomas and Jones 1993). This occurred because the processes causing their declines were suddenly reversed (section 8.4). For this and other reasons, we believe that predictions of extinction are likely to be accurate only if they are based on an understanding of the underlying mechanisms.

8.3   Patterns among declining species

General

Probably many different factors have caused declines among invertebrates, given the diversity of change in the British countryside (Rackham 1986): more than 50 different threats are listed in the Red Data Books (RDBs). However, autecological studies and reviews of dragonflies (Moore 1976; Mendel 1992), butterflies (Thomas 1984, 1991), and beetles of woodland (Warren and Key 1991) and fresh water (Foster 1991, 1992) suggest that some factors are particularly important and that taxa with certain lifestyles or habitat-types may be especially threatened. We therefore analysed the information in RDBs and other reviews to seek patterns among declining species, concentrating on their habitats, because nearly every species account implies some sort of habitat change as the main threat. Interestingly, collectors were implicated in the declines of just two species.

We found no clear pattern in the food or trophic level of extinct and RDB species, except that specialists are undoubtedly more endangered that generalists. Phytophagous taxa, pollen-feeding Hymenoptera, and aquatic and terrestrial carnivores have a similar representation (Table 8.1). Shaw (1987) plausibly suggests that parasitic Hymenoptera have experienced greater declines than other insects due to their extreme trophic position, but the data are too poor to demonstrate this. He also describes parasitoids that have declined greatly when their hosts have not, analogous to the numerous local extinctions of butterfly populations on sites where their larval foodplants have remained abundant (Thomas 1991). Many aculeate Hymenoptera live at intermediate trophic levels as brood parasites (usually of one or a few host species), cleptoparasites, and inquilines. This might partly explain why a higher proportion of this group is listed as both RDB and Notable species than any other (Falk 1991). However, most also inhabit the two most threatened types of terrestrial habitat in Britain (see below). Myrmecophily is another common trait among RDB species, but too little is known about the life histories of most insects to determine whether myrmecophilous species are especially threatened: obligate myrmemophiles undoubtedly are (Erhardt and Thomas 1991).

The comparative vulnerability of dispersive and sedentary species, and of those living near the edges of their range in Britain, are discussed in section 8.4. We also examined size as a factor in butterflies, but found no relationship between changes in status this century and wing size.

The biotopes and habitats of Red Data Book species

Welch (1993) classified the biotopes of all RDB and notable species of British invertebrates (Fig. 8.2), and obtained similar patterns for 'endangered' and 'vulnerable' species alone. Most inhabit deciduous woodland, coastal, and aquatic (freshwater) biotopes or heathland, whereas upland, calcareous, and other grasslands contain comparatively few. The low 'score' of calcareous grassland is surprising given its emphasis by conservationists. However, these figures show where the most threatened species live rather than the invertebrate fauna as a whole.

Welch also found considerable variation in the occupancy of these biotopes by threatened species in different taxonomic groups (Fig. 8.3). Deciduous woodland proved particularly important for Diptera and Coleoptera, heathland for aculeate Hymenoptera and spiders, and aquatic biotopes for snails.

In a review of European butterfly conservation, Thomas (1991) distinguished between extinctions caused by the fundamental destruction or fragmentation of biotopes (e.g., conversion of heathland to agriculture) and those caused by the less obvious disappearance of species habitats within surviving biotopes, usually through successional change. For western European butterflies, the second process has caused many more extinctions in recent decades and autecological

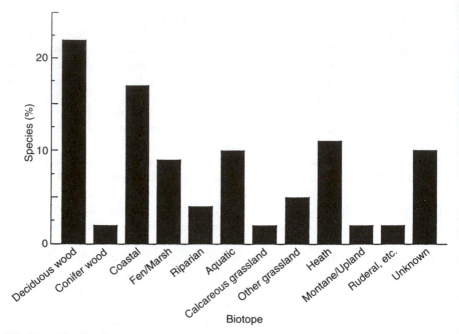

Fig. 8.2   The distribution of all RDB and Notable invertebrates ($N = 2518$) in 12 principal types of biotope. (After Welch 1993.)

studies of other endangered terrestrial invertebrates support this conclusion (e.g., Cherrill and Brown 1990a,b, 1992; L. K. Ward, pers. comm.). The pattern of threatened invertebrates in biotopes (Fig. 8.2) suggests this may apply to other taxa. Certainly, there is no correlation between the extent by which biotopes have declined in historical times and the number of threatened species they contain; indeed, greater proportions of deciduous woodland, heathland, and most coastal biotopes have survived than calcareous and other lowland grasslands, fens, and marshes (Warren 1992), yet the first three biotopes possess more threatened and extinct species than the latter group (rank correlation $r = 0.60$, $N = 5$).

We therefore classified all 'extinct', 'endangered', and 'vulnerable' terrestrial invertebrates by the type of habitat they possess rather than by the biotope they inhabit. Where possible, species were assigned to one of six successional stages: (1) bare ground and very early, including dunes but not shingle, pioneer heath, woodland floors in the first 2 years after coppicing or clearing, grassland within 2 years of perturburation or maintained as a plagioclimax below 3 cm tall: (2) the building phase of heaths, 5–10 year regrowth after coppicing or woodland clearing, 4–10 cm tall grassland; (3) mature heath, grassland above 10 cm tall, including woodland glades and rides; (4) shrubs; (5) healthy trees; (6) saproxylic, defined as species that inhabit dead or dying, standing or fallen wood, or the fungi growing on it, or species supported by other saproxylic species (Speight 1989), such as the mymecophilous Coleoptera of *Lasius brunneus* nests. Seven experts

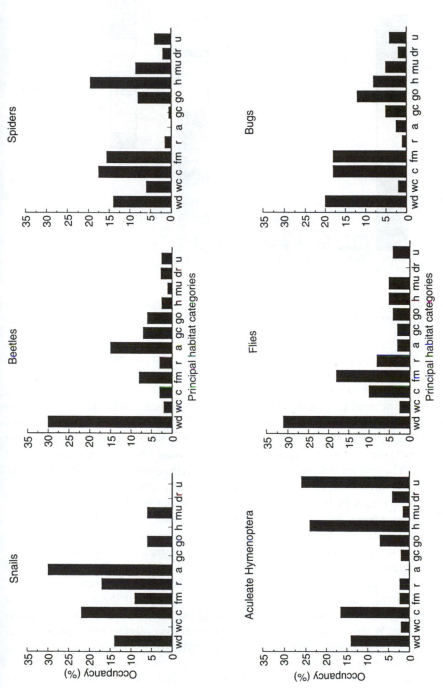

Fig. 8.3 The representation of 6 taxa of RDB and Notable invertebrates in 12 principal types of British biotope. (After Welch 1993.)

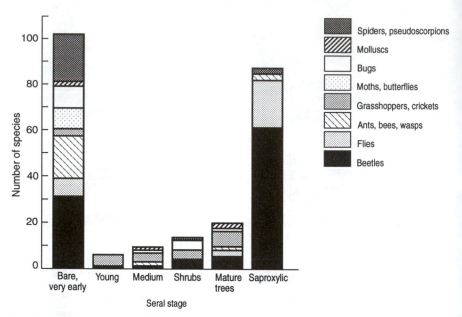

Fig. 8.4 The seral stages inhabited by 'extinct', 'endangered', and 'vulnerable' RDB species in woods, grassland, heaths, and dunes (a further 65 species could not be classified).

helped us categorize their specialist groups. All mymecophilous species characteristic of woodland clearings (apart from those associated with *L. brunneus*) were assigned to category (1); where two categories were used during a life cycle, the seral stage needed by early stages was preferred, because they have more exacting requirements than adults (N. W. Moore, pers. comm.; Thomas 1991). Exceptions, such as *Decticus verrucivorus* which requires a mosaic of category (1) and (3) seral stages (Cherrill and Brown 1990*b*), were scored as half in each. The one bumble bee listed was tentatively scored in category (3) due to the critical adult need in this group for abundant nectar.

Forty-five species had unknown or very specialized habitats, such as the Coleoptera known only from osprey or hornet nests, squirrel drays, and dung, and inhabitants of cliffs and quarries (possibly category 1). We had insufficient information to assign a further 65 species to such precise seral stages. However, the remaining 232 species revealed a remarkably clear pattern, with the great majority shown as inhabitants of very early or very late seral stages (Fig. 8.4). This confirms, quantifies, and extends the hypothesis that the most threatened groups of woodland insect in Britain depend on the extremes of the woodland succession (Warren and Key 1991). The species that require early successional stages are taxonomically varied, but the saproxylic group consists almost entirely of Coleoptera and Diptera (Fig. 8.4).

We have no comparable figures to show where all British invertebrates live, but the pattern is probably a rough mirror image of Fig. 8.4 (although less extreme), with fewest species in the earliest successional stages and most inhabiting mature trees. The dependence of threatened species on the extremes of seres is discussed in section 8.4.

8.4   Mechanisms causing declines

Terrestrial species

Early successional and thermophilous invertebrates

The British niches of 12 species studied that inhabit the earliest stages of seres (1 ant, 1 orthopteran, 10 butterflies) are even narrower than described, because all are also restricted to south-facing slopes and most to southern England (Cherrill and Brown 1990a,b, 1992; Thomas 1984, 1991, 1993; Warren 1994). These species do not need early successions *per se*, merely the exceptionally warm microclimates associated with these situations in which to develop (Thomas 1983b, 1993; Cherrill and Brown 1990b; Warren 1994). Thus, under the warmer climate of lowland central France, the same species occur on all aspects of land *except* south-facing slopes (which are too hot) and in tall rather than short swards or in semi-shaded woodland (Thomas 1993). This may not apply to all the other 98 species in this category (Fig. 8.4), but as many reach the northern limit of their ranges in southern England, most are probably thermophilous, ground-dwelling species that can inhabit only the hottest patches available here.

The constraint caused by the extra requirement of an unusually warm local climate has been calculated for the butterfly *Plebejus argus* using the Dorset heathlands database and descriptions of its habitat from the extreme north of its range and from further south, where summer temperatures are 2 °C warmer (Thomas 1991). Under the cooler climate, only 7 small islands of *P. argus* habitat would exist in just 4 heaths, whereas under a warmer climate, precisely the same biotopes would contain a 70-times greater area of habitat distributed as 112 islands within 45 sites (Fig. 8.5). The probability of local extinction is clearly greater near the edge of *P. argus*' range and the probability of recolonization is lower due to the greater distance between colonies. Moreover, butterfly populations near the edge of species' ranges undergo much greater fluctuations than those further south, further increasing the chance of local extinction in unfavourable years (Thomas *et al.* 1994).

In fact, thermophilous species have survived in Britain for long periods on a few restricted sites, provided that their specialized habitats were maintained: *Melitaea cinxia* populations have bred on the crumbling undercliffs of the Isle of Wight for at least 170 generations (Heath *et al.* 1984). But this butterfly is unusual. Most declining species of early seral stages have depended entirely on man (or rabbits) to create their ephemeral habitats since historical records began, 150–200 years ago (e.g., Thomas 1993). Some may be relatively recent colonists from southern Europe, but the low vagility, distributions or subspeciation of others suggest that they descend from populations which colonized the country in 9000–7000 BP (before present), before the land link with the continent was broken (Dennis 1992). During that period, mean summer temperatures were

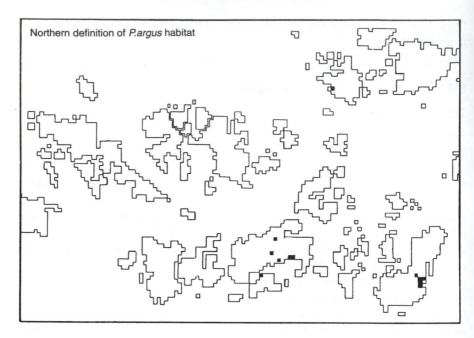

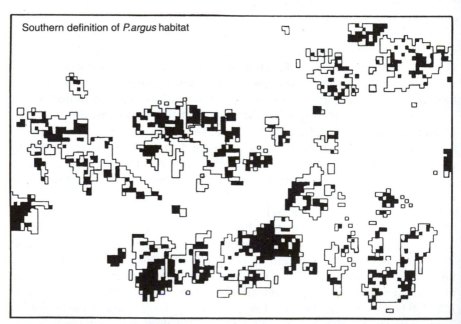

Fig. 8.5 The distribution of the habitat of *Plebejus argus* on Dorset heathland in 1978, using habitat definitions made at the northern edge of its range (top) and further south where the mean summer climate is about 2 °C warmer (bottom). Solid lines indicate boundaries of heaths (=distribution of *P. argus* foodplants); ■, distribution of *P. argus* habitat. (From Thomas 1991.)

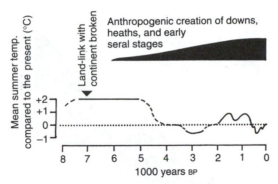

Fig. 8.6   Summer temperatures and the availability of warm habitats in Britain. (From Thomas 1993.)

about 2 °C warmer than today (Fig. 8.6), so their habitats would have been much more common (Fig. 8.5), less ephemeral, and would not have required man to maintain them. Thomas (1993) suggests that these ground-dwelling species did not retreat south when the summer climate cooled, around 5000 BP, because by then man had been creating warm habitats within woods, heath, and grassland for at least 1000 years (Fig. 8.6), providing refuges where thermophilous species could survive.

Whatever the merits of this hypothesis, archaeology and history indicate that man's traditional forms of land management in Britain resulted in the frequent and regular regeneration of early successional stages—or the maintenance of short plagioclimaxes—on an increasingly large scale within woods, heaths, and grasslands from about 6000 BP until the late 19th century (Rackham 1986). But in the 20th century, there has been a shift towards later seral stages as coppicing was replaced by high forest (Fig. 8.7), and heaths and unfertilized grassland abandoned (Smith 1980; Webb 1986). Thermophilous butterflies, and presumably other taxa, experienced numerous local extinctions as their southern habitats became shadier; these butterflies experienced cooling (-4 °C) equivalent to transportation to the Orkneys (Thomas 1983b, in press; Thomas *et al.* 1986). These changes were quite independent of the survival of biotopes. Woodland of all types has doubled during this period, but early successional habitats within woods became rare and isolated (Fig. 8.7).

Saproxylic species

The threatened invertebrates of dead or dying wood are difficult to study and few autecological studies have been made. Reviews by Speight (1989) and Warren and Key (1991) are sufficiently thorough for us to comment only briefly on some similarities and differences between these species and those inhabiting early stages of seres.

One similarity is that most species of this stage are specialists that inhabit extremely narrow and specific niches (Warren and Key 1991). These have been

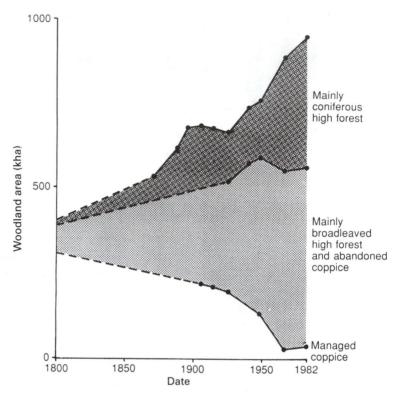

Fig. 8.7   Changes in the area and character of woodland biotope in England since 1800. (From Warren and Key 1991.)

categorized into 20 broad types. So specialized is the habitat of some species that only about 1% of ancient trees may contain it at any one time (Stubbs 1972).

Microclimate is important in determining whether dead wood will develop a species' habitat (Warren and Key 1991), but unlike the early successional insects, most saproxylics cannot live in hot microclimates. Many require very wet conditions and some only inhabit wood that has fallen into water; others occur in arid red-rotted wood or fire-scorched timber.

Warren and Key (1991) found no consistent patterns of extinction among saproxylic species, except that Caledonian pine forest species have declined particularly severely in historical times, whereas those associated with decaying exotic softwoods have increased. Declines presumably started with the major forest clearances around 4000 BP, which reached a maximum in the early 19th century. Since then the woodland biotope has doubled (Fig. 8.7), but most modern woods are middle-aged, managed, and contain very little dead woods. Saproxylic habitats have also declined in ancient parks and forests, for moribund trees are regarded as untidy, dangerous to the public, and (wrongly) as a source of infection to sound trees.

Wetland and freshwater habitats

Perceived threats and causes of loss to the large number of threatened and extinct invertebrate species associated with various wetland biotopes (Figs 8.2 and 8.3) are difficult to interpret, with several factors sometimes listed for the same species. This may genuinely reflect the diversity of factors harming wetland habitats in the British countryside, but more research is needed to clarify this: only information of water beetles (Foster 1991, 1992) and Odonata (e.g., Moore 1976) is reasonably clear. Foster (1991) reviews the numerous perceived threats listed in RDBs. Our figures, from slightly different sources, are very similar: the main factors blamed are drainage, including the lowering of water tables and the loss of temporary ponds (62 citations), pollution and/or eutrophication (26 citations), and inappropriate management (13 citations).

Fragmentation and dispersal

Little is known about the mobility of threatened invertebrates. British butterflies can be divided crudely into mobile species with open populations (about 15%) and those that are colonial (85%) (Thomas 1984). No pattern is discernable in the mobility of the few extinct species: *Aporia crataegi* and *Nymphalis polychloros* lived in predominantly open populations (Heath *et al.* 1984); *Carterocephalus palaemon* (extinct in England) is also comparatively mobile (Ravenscroft 1994), but three other extinct butterflies lived in closed populations and *M. arion* was highly sedentary (Thomas 1994). The other butterfly species that declined severely this century appear to be sedentary, and have often failed to track the rate at which their habitats have been created, most being incapable of crossing barriers of > 1–10 km (Thomas 1991; Thomas *et al.* 1992; Thomas and Jones 1993; Thomas 1994).

There is anecdotal evidence that many other declining invertebrates possess sedentary populations, especially saproxylic species (Warren and Key 1991). This is expected from templet theory (Southwood 1977). However, the very low vagility of British species with ephemeral early successional habitats is surprising. It is possible that, having become trapped in anthropogenic habitats after the climate cooled (Thomas 1993), these ground-dwelling thermophilous arthropods evolved local races adapted to the availability of their habitat patches under traditional forms of land management (Thomas 1991). For example, low vagility may have been advantageous to the butterfly *Boloria euphrosyne* when it bred in freshly cleared coppice panels, because under coppicing its habitat was created about nine times more frequently and patches existed nine times nearer to one another than is the case in modern woodland; in addition, the dynamics of at least three of Southwood's six other components of a species' habitat would have favoured sedentariness under coppice management (Thomas 1991). Dempster (1991) suggests that reduced vagility may have evolved in as few as 30 generations in two British butterfly species when the availability of their habitats changed. It would be surprising, therefore, if populations that were exposed for 2000–5000 generations to the artificial patch dynamics of traditional forms of land managements had not experienced selection. But if this is so, conservationists face formidable problems in regions where there is a long history of now-obsolete land management.

8.5 Conclusions

Many invertebrate species are declining at rates that match and probably exceed those of vertebrates and vascular plants in Britain. Although few national extinctions have been recorded, many species are on the verge of extinction. A similar or worse situation exists in the Netherlands (van Swaay 1990) and parts of Germany (O. Kudrna, pers. comm.). More thorough scientific recording is essential if declines are to be recognized: a shortage of taxonomists is a major problem in Britain.

It is important to understand the underlying mechanisms if successful predictions are to be made about how invertebrate populations will change in modern landscapes and, probably, under changing climates. The preponderance of thermophilous species among threatened invertebrates was a surprise; with the prospect of global warming, conservationists should nurture these populations, for they may multiply in the future.

The extent to which man has altered both the biotopes of invertebrates and species habitats within them is clear. In Britain, species of the two extremes of succession are especially threatened. These types of habitat are often scarce on nature reserves. Management should create (or leave) more of them, and new habitats should be close enough in space and time to be utilized by sedentary species. These same habitat-types have become rare at similar latitudes across Europe, but elsewhere saproxylic species form a much more threatened group, as is probably the case in other continents. Conservationists outside Europe should also consider the possibility that populations of some invertebrate species may have become confined (and perhaps adapted) to antique anthropogenic habitats that were maintained by prehistoric forms of land management; the dependence of the North American butterfly *Speyeria zerene hippolyta* on the frequent grass fires set by Indians before European settlement is one example (Hammond and McCorkle 1984).

The exact requirements of only a few representative invertebrate species can be studied, but more autecological studies are required from saproxylic and aquatic taxa, which may reveal clearer patterns about causes of decline. Although Prendergast *et al.* (1993) found little overlap in the 'hotspots' where the diversity of different taxa was greatest in Britain, greater understanding may yet enable us to make more generalizations.

Finally, the lack of knowledge of dispersal by endangered invertebrates is a major constraint for conservationists trying to apply metapopulation concepts to their conservation (e.g., Thomas 1994). This will be particularly important if there is a redistribution of invertebrates following climate change. But even under the status quo, the practical questions of how often, how large, and how far apart species habitat patches should be created remain largely unanswered.

Summary

This chapter reviews what is known about the changing status and extinction rates of invertebrates in Britain during the past 100–300 years. Although historical recording was more thorough in Britain than elsewhere, the data are patchy and difficult to interpret. Nevertheless, we conclude that within Britain, the extinction rates of invertebrates have matched, and probably exceeded, those of vertebrates and vascular plants in the present century. The main reasons for decline are analysed. No clear pattern was found among the numerous threatened species in aquatic biotopes, but there was a very clear pattern in terrestrial biotopes, where most threatened species inhabit either very early or very late successional stages. The former group consists mainly of thermophilous species, which might be relics from a period when British summer temperatures were warmer, and which survived because prehistoric man created warm refugia by traditional forms of land management. Not only have these types of habitat largely disappeared from modern biotopes, but their dynamics have also changed. Some invertebrates are too sedentary to track their habitats in the modern landscape. We emphasize that the focus in this chapter is on extinction within Britain; none of these species has become globally extinct—yet.

Acknowledgements

We are very grateful to H. Arnold, S. Ball, G. W. Elmes, G. R. Else, B. C. Eversham, S. Corbett, P. T. Harding, R. S. Key, J. H. Lawton, R. M. May, H. Mendel, N. W. Moore, R. G. Snazell, L. K. Ward, C. D. Thomas, L. K. Ward, M. S. Warren, R. C. Welch, P. H. Williams, and I. Woiwod for comments or advice; to S. Creer for assistance with illustrations; and to R. C. Welch and Her Majesty's Inspectorate of Pollution for permission to reproduce Figs 8.2 and 8.3 from Welch (1993).

References

Alexander, K. N. A. (1992). The work of the National Trust biological survey team. In *Biological recording of changes in British wildlife*, (ed. P. T. Harding), p. 73. HMSO, London.

Allan, P. B. M. (1943). *Talking of moths*. Montgomery, Newtown.

Beek, T. P., Ovaa, A. H., and van der Made, J. G. (ed.) (1992). *Future of butterflies in Europe*. Wageningen Agricultural University, Wageningen.

Bratton, J. H. (1990). *A review of the scarcer Ephemoroptera and Plecoptera of Great Britain*. Nature Conservancy Council, Peterborough.

Bratton, J. H. (ed.) (1991). *British Red Data Books, 3. Invertebrates other than insects*. Joint Nature Conservation Committee, Peterborough.

Cherrill, A. J. and Brown, V. K. (1990a). The life cycle and distribution of the wart-biter

Decticus verrucivorus (L.) (Orthoptera: Tettigoniidae) in a chalk grassland in southern England. *Biol. Conserv.*, **53**, 125–43.

Cherrill, A. J. and Brown, V. K. (1990*b*). The habitat requirements of adults of the wart-biter *Decticus verrucivorus* (L.) (Orthoptera: Tettigoniidae) in southern England. *Biol. Conserv.*, **53**, 145–7.

Cherrill, A. J. and Brown, V. K. (1992). Ontogenetic changes in the micro-habitat preferences of *Decticus verrucivorus* (Orthoptera: Tettigoniidae) at the edge of its range. *Ecography*, **15**, 37–44.

Collins, N. M. and Thomas, J. A. (ed.) (1991). *The conservation of insects and their habitats*. Academic Press, London.

Dempster, J. P. (1991). Fragmentation, isolation and mobility of insect populations. In *The conservation of insects and their habitats*, (ed. N. M. Collins and J. A. Thomas), pp. 143–54. Academic Press, London.

Dennis, R. L. H. (ed.) (1992). *The ecology of butterflies in Britain*. Oxford University Press.

Duffey, E. and Watt, A. S. (ed.) (1971). *The scientific management of animal and plant communities for conservation*. Blackwell, Oxford.

Erhardt, E. and Thomas, J. A. (1991). Lepidoptera as indicators of change in semi-natural grasslands of lowland and upland Europe. In *The conservation of insects and their habitats*, (ed. N. M. Collins and J. A. Thomas), pp. 213–36. Academic Press, London.

Falk, S. (1991). *A review of the scarce and threatened bees, wasps and ants of Great Britain*. Nature Conservancy Council, Peterborough.

Ford, E. B. (1945). *Butterflies*. Collins, London.

Foster, G. N. (1991). Conserving insects of aquatic and wetland habitats, with special reference to beetles. In *The conservation of insects and their habitats*, (ed. N. M. Collins and J. A. Thomas), pp. 237–62. Academic Press, London.

Foster, G. N. (1992). The effects of changes in land use on water beetles. In *Biological recording of changes in British wildlife*, (ed. P. T. Harding), pp. 27–30. HMSO, London.

Hammond, P. C. and McCorkle, D. V. (1984). The decline and extinction of *Speyreia* populations resulting from human environmental disturbances. *J. Res. Lepid.*, **22**, 217–24.

Harding, P. T. (1992). *Biological recording of changes in British wildlife*. HMSO, London.

Harding, P. T. and Sheail, J. (1992). The Biological Records Centre—a pioneer in data gathering and retrieval. In *Biological recording of changes in British wildlife*, (ed. P. T. Harding), pp. 5–19. HMSO, London.

Heath, J. (1981). *Threatened Rhopalocera (butterflies) in Europe*. Council of Europe, Strasburg.

Heath, J., Pollard, E., and Thomas, J. A. (1984). *Atlas of butterflies in Britain and Ireland*. Viking, Rickmansworth.

Hyman, P. S. and Parsons, M. S. (1992). *A review of the scarce and threatened Coleoptera of Great Britain*, Part 1. Joint Nature Conservation Committee, Peterborough.

Kerney, M. P. (1976). *Atlas of the non-marine Mollusca of the British Isles*. Institute of Terrestial Ecology, Cambridge.

Kerrich, G. J., Hawksworth, D. L. and Sims, R. W. (1978). *Key works to the fauna and flora of the British Isles and northwest Europe*. Academic Press, London.

Kirby, P. (1992). *A review of the scarce and threatened Hemiptera of great Britain*. Joint Nature Conservation Committee, Peterborough.

Marshall, J. A. and Haes, E. C. M. (1988). *Grasshoppers and allied insects of Great Britain and Ireland*. Harley, Colchester.

Mendel, H. (1992). *Suffolk dragonflies*. Suffolk Naturalist's Society, Ipswich.

Merrett, P. (1990). *A review of the nationally notable spiders of Great Britain*. Nature Conservancy Council, Peterborough.

Moore, N. W. (1976). The conservation of Odonata in Great Britain. *Odonatologica*, **5**, 37–44.

Munguira, M. L. and Thomas, J. A. (1992). The use of road verges by butterfly and burnet populations, and the effect of roads on adult dispersal and mortality. *J. Appl. Ecol.*, **29**, 316–29.

Parsons, M. S. (1993). *A review of the scarce and threatened pyralid moths of Great Britain*. Joint Nature Conservation Committee, Peterborough.

Perring, F. H. (1992). BSBI distribution maps scheme—the first 40 years. In *Biological recording of changes in British wildlife*, (ed. P. T. Harding), pp. 1–4. HMSO, London.

Pollard, E. and Yates, T. J. (1993). *Monitoring butterflies for ecology and conservation*. Chapman & Hall, London.

Prendergast, J. R., Quinn, R. M., Lawton, J. H., Eversham, B. C., and Gibbons, D. W. (1993). Rare species, the coincidence of diversity hotspots and conservation strategies. *Nature*, **365**, 335–7.

Rackham, O. (1986). *The history of the countryside*. Dent, London.

Ravenscroft, N. O. M. (1994). The conservation of the chequered skipper butterfly in Scotland. In *Ecology and conservation of butterflies*, (ed. A. S. Pullin), pp. 160–79. Chapman & Hall, London.

Shaw, M. R. (1987). Hymenoptera: Parasitica. In *British Red Data Books*, 2. *Insects*, (ed. D. B. Shirt), pp. 257–8. Nature Conservancy Council, Peterborough.

Shirt, D. B. (ed.) (1987). *British Red Data Books* 2. *Insects*. Joint Nature Conservation Council, Peterborough.

Smith, C. J. (1980). *Ecology of the English chalk*. Academic Press, London.

South, A. (1974). Changes in the composition of the terrestrial mollusc fauna. In *The changing flora and fauna of Britain*, (ed. D. L. Hawksworth), pp. 255–74. Academic Press, London.

Southwood, T. R. E. (1977). Habitat, the templet for ecological strategies? *J. Anim. Ecol.*, **46**, 337–65.

Speight, M. C. D. (1989). *Saproxylic invertebrates and their conservation*. Council of Europe, Strasbourg.

Spellerberg, I. F., Goldsmith, F. B., and Morris, M. G. (1991). *The scientific management of temperate communities for conservation*. Blackwell, Oxford.

Stubbs, A. E. (1972). Wildlife conservation and dead wood. Supplement to *J. Devon Trust Nat. Conserv.*, 1–18.

Tax, M. R. (1989). *Atlas van de nederlandse dagvlinders*. 's-Graveland, Wageningen.

Thomas, C. D. (1994) Extinction, colonization and metapopulations: environmental tracking by rare species. *Conserv. Biol.*, **8**, 373–8.

Thomas, C. D. and Jones, T. M. (1993). Partial recovery of a skipper butterfly (*Hesperia comma*) from population refuges: lessons for conservation in a fragmented landscape. *J. Anim. Ecol.*, **62**, 472–81.

Thomas, C. D., Thomas, J. A., and Warren, M. S. (1992). Distributions of occupied and vacant butterfly habitats in fragmented landscapes. *Oecologia*, **92**, 563–7.

Thomas, J. A. (1983a). A quick method for estimating butterfly numbers during surveys. *Biol. Conserv.*, **27**, 195–211.

Thomas, J. A. (1983b). The ecology and conservation of *Lysandra bellargus* (Lepidoptera: Lycaenidae) in Britain. *J. Appl. Ecol.*, **20**, 59–83.

Thomas, J. A. (1984). The conservation of butterflies in temperate countries: past efforts and lessons for the future. In *Biology of butterflies*, (ed. R. Vane-Wright and P. Ackery), pp. 333–53. Academic Press, London.

Thomas, J. A. (1991). Rare species conservation: case studies of European butterflies. In

The scientific management of temperate communities for conservation, (ed. I. Spellerberg, B. Goldsmith, and M. G. Morris), pp. 149–97. Blackwell, Oxford.

Thomas, J. A. (1993). Holocene climate change and warm man-made refugia may explain why a sixth of British butterflies inhabit unnatural early-successional habitats. *Ecography*, **16**, 278–84.

Thomas, J. A. (1994). The ecology and conservation of *Maculinea arion* and other species of large blue. In *Ecology and conservation of butterflies*, (ed. A. S. Pullin), pp. 180–97. Chapman & Hall, London.

Thomas, J. A., Moss, D., and Pollard, E. (1994). Increased fluctuations of butterfly populations towards the northern edges of species' ranges. *Ecography*, **17**, in press.

Thomas, J. A., Thomas, C. D., Simcox, D. J. and Clarke, R. T. (1986). The ecology and declining status of the silver-spotted skipper butterfly (*Hesperia comma*) in Britain. *J. Appl. Ecol.*, **23**, 365–80.

Tutt, J. W. (1905–14). *A natural history of the British butterflies*, Vols 1–4. London.

van Swaay, C. A. M. (1990). An assessment of the changes in butterfly abundance in the Netherlands during the 20th century. *Biol. Conserv.*, **52**, 287–302.

Wallace, I. D. (1991). *A review of the Trichoptera of Great Britain*. Nature Conservancy Council, Peterborough.

Warren, M. S. (1992). The conservation of British butterflies. In *The ecology of butterflies in Britain*, (ed. R. L. H. Dennis), pp. 246–74. Oxford University Press.

Warren, M. S. (1994). Managing local microclimates for the high brown fritillary. In *Ecology and conservation of butterflies*, (ed. A. S. Pullin), pp. 198–215. Chapman & Hall, London.

Warren, M. S. and Key, R. S. (1991). Woodlands: past, present and potential. In *The conservation of insects and their habitats*, (ed. N. M. Collins and J. A. Thomas), pp. 155–212. Academic Press, London.

Warren, M. S., Thomas, C. D., and Thomas, J. A. (1984). The status of the heath fritillary butterfly *Mellicta athalia* Rott. in Britain. *Biol. Conserv.*, **29**, 287–305.

Webb, N. R. (1986). *Heathlands*. Collins, London.

Welch, R. C. (1993). *An assessment of the invertebrates associated with the Institute of Terrestrial Ecology's land cover classes*. Institute of Terrestrial Ecology, Cambridge.

Woiwod, I. P. and Harrington R. (1994). Flying in the face of change; the Rothamsted insect survey. In *Long-term experiments in agriculture and ecological sciences*, (ed. R. A. Leigh and A. E. Johnston), pp. 321–41. CAB International, Wallingford, UK.

9

Assessing the risk of plant extinction due to pollinator and disperser failure

W. J. Bond

What escapes the eye, however, is a much more insidious kind of extinction: the extinction of ecological interactions.

(Janzen 1974)

9.1 Introduction

Ecology has contributed two major insights to the biology of extinction: large areas hold more species than smaller areas and larger populations persist longer than smaller ones. No comparable generalization has emerged from studies of ecological interactions. In this chapter I attempt a new approach to predicting extinction which explicitly includes the importance of interactions. The patterns that emerge are not used to make general statements on probable species loss but rather to identify general traits which increase extinction risk. I use the interaction between plants and their pollinators and dispersers as an example of this species-centred view. These reproductive mutualisms epitomize the subtle, complex web of interactions which, if broken by human actions, could cause a cascade of extinctions. Janzen (1974, 1987) has eloquently drawn attention to the possible 'extinction of ecological interactions' in tropical forests as a result of habitat transformation. Similar concerns, focusing on pollination, have been raised in other parts of the world including pesticide-sprayed north-temperate woodlands (Kevan 1975), alpine meadows (Vogel and Westerkamp 1991), and eroded African range lands (F. W. Gess and S. K. Gess 1993).

To address the question of whether a wave of plant extinctions will follow the collapse of pollinator and dispersal interactions, and to help avert them, I ask:

(1) Which traits indicate vulnerability?

(2) Which species are threatened and in what systems?

There is an enormous literature on pollination and a growing literature on dispersal (see e.g., Boucher 1985; Estrada and Fleming 1986) very little of which is informative for predicting extinction (although see Gilbert 1980; Howe 1984; Addicott 1986). The biology of reproductive mutualism is notoriously complex (Boucher 1985; Howe and Westerley 1988). To make the problem of predicting extinctions manageable, I focus on ecological rather than genetic or evolutionary consequences and ecological timescales of centuries rather than evolutionary scales of millennia.

9.2 Threats to pollinators and dispersers, and plant responses

Pollinators and dispersers, both vertebrate and invertebrate, face diverse threats. These include poisoning by pesticides (Johannsen 1977; Kevan 1975; Kevan *et al.* 1985), habitat alteration (Janzen 1987; F. W. Gess and S. K. Gess 1993), invasions of alien animals and plants (Bond and Slingsby 1984; Breytenbach 1986), and insularization of habitats (Linhart and Feinsinger 1980; Diamond 1984; Jennersten 1988). There seems little doubt that the community of pollinators and dispersers is being altered by these and other forces. The consequences are likely to be a reduction in diversity of animal mutualists, and changes in population densities of the survivors. It is tempting to argue that specialists dependent on a few species will be more vulnerable than generalists. However, the diversity of threats is so great that whole assemblages of mutualists may be eliminated. No mutualism is completely assured.

The pathway to extinction is not a simple one for plants deprived of their mutualist partners. Key considerations are: (1) the probability of a mutualism failing due to the demise of mutualist partners; (2) the degree of reproductive dependence on mutualism (facultative or obligate); and (3) the importance of seeds in the demography of the plant. All are important in assessing extinction risk. For example, inclement weather often disrupts pollinating insects in temperate and alpine habitats. However, many herbaceous plants in these habitats are self-pollinated or have such a diverse pollinator fauna that seed set is assured. A few with specialist pollinators may fail to set seed but persist by vegetative propagation. These compensatory mechanisms—self-pollination, flowers pollinated by a generalist fauna, or escape from demographic dependence on seeds—may ensure persistence in many habitats where pollinator services are unreliable. Similar considerations apply to dispersal mutualisms. Each of the contributing factors can be evaluated on a scale indicating increasing vulnerability to provide simple qualitative criteria for assessing risk. Far more complex approaches are needed for fully quantitative analyses of particular populations (e.g., Price and Jenkins 1986).

9.3 The probability of mutualisms failing

Both the likelihood of a particular mutualist being lost and the possibility of its reproductive services being replaced need consideration. Most plants are pollinated by several to many species, often of widely diverse taxonomic origin (Fig. 9.1; see e.g., Feinsinger 1983, 1987; Janzen 1983; Schemske 1983; Herrera 1984, 1988). Pollination by a single species seems to be very rare although figs, *Yucca*, and orchids include notable exceptions (Fig. 9.1). If field studies are

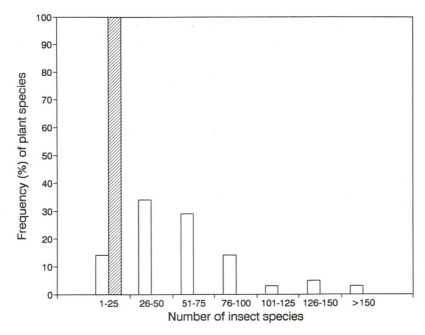

Fig. 9.1 Frequency distribution of insect visitors to 55 species of sunflowers in Illinois (open bars; ex Schemske 1983) contrasted with confirmed insect pollinators of 36 Australian orchid species (hatched bars; ex Adams and Lawson 1993). The diversity of flower visitors can vary greatly among geographical regions or different floral types.

lacking, pollinator specificity may be crudely assessed from floral morphology. Increasing specialization is associated with traits that limit pollinator access including complex shapes, large size, long corolla tubes, and floral orientation (Faegri and van der Pijl 1979; Linhart and Feinsinger 1980; Motten 1986) or specialized attractants and rewards (Dafni *et al.* 1990; Johnson and Bond 1994; Steiner 1989). Dafni (1992) provides a useful practical guide to pollination biology.

Increasing specificity of dispersal agents is associated with increasing fruit or seed size for birds (Martin 1985; Herrera 1984; Wheelwright 1985) and some mammal faunas (Janzen 1986; Janzen and Martin 1982). Similar trends have been reported for Australian ant-dispersed species where large seeds with large rewards are more likely to be carried by non-destructive ants (Hughes and Westoby 1992). Jordano (1992) has provided a useful review of vertebrate frugivory which considers both traits associated with selectivity and the demographic consequences of frugivory.

Using these criteria or others derived from field knowledge, the probability of pollinator or disperser failure for a plant species can be ranked as illustrated in Table 9.1. I have used an arbitrary scale from 0 (wind or abiotic) to 1 (single-species dependence). Plants served by several closely related mutualist species may be more vulnerable than those served by a taxonomically diverse fauna.

Table 9.1 Plant attributes and extinction risk. Extinction risk is greatest when the risk of pollinator/dispersal failure, reproductive dependence on the process and demographic dependence on seeds are all high.

| Rank | Risk of process failure (*PS, DS*) | Dependence on process | | Dependence on seeds (*SD*) |
		Pollination (*BS*)	Dispersal (*DD*)	
High (1)	Single-species dependence	Dioecious Self-incompatible	Dispersal obligatory to cue germination, reach safe sites, evade predators	Seed propagation only, lifespan (10^0–10^2 y), killed by disturbance, few large seeds, no seed bank, sparse seedlings
	Specialist Generalist	Self-compatible		
Low (0)	Wind	Self-pollinated apomicts	Dispersal not needed for germination, recruitment or survival	Vegetative propagation, lifespan (10^2–10^4 y), resprouts after disturbance, many small seeds, persistent seed bank, dense seedlings

PS, pollinator; *DS*, disperser specificity; *BS*, breeding system; *DD*, dispersal dependence; *SD*, population dependence on seeds

9.4   Reproductive dependence on mutualism

The degree of dependence on mutualism for reproductive services varies greatly. The breeding system of a plant is of major importance in assessing its dependence on pollinators. This may vary from none in asexual and self-pollinated species to facultative in self-compatible species and obligate in self-incompatible and dioecious plants (Richards 1986). Rapid surveys of breeding systems can be made from herbarium material (e.g., Cruden 1977; Plitmann and Levin 1990) but should preferably be supplemented by hand-pollination studies in the field (e.g., Bawa et al. 1985b; Dafni 1992). The frequency of asexual seed production may be underestimated by these simple methods (see e.g., Ha et al. 1988 for tropical forest trees). Breeding systems can be ranked on a scale from 0 to 1 to indicate relative dependence on pollination for seed set (Table 9.1).

The dependence on dispersal for seedling recruitment is far less well understood but it appears to vary just as much with some species highly dependent on dispersers for germination (e.g., Janzen 1983 for a review) or recruitment (e.g., Augspurger 1984; Beattie 1985; Slingsby and Bond 1985; Louda 1989; Jordano 1992), and others not (e.g., Janzen 1983; Pierce and

Cowling 1991). Fugitive species, occupying ephemeral habitats, may be particularly dependent on dispersal, especially where they lack seed banks (Louda 1989) in contrast to trees that form seedling banks and tolerate shaded conditions until openings occur. But in principle, dispersal dependence too can be ranked on a scale of 0 to 1 indicating increasing dependence on the process for recruitment (Table 9.1).

9.5   Demographic dependence on seeds

Reproductive mutualisms are a favourite subject for the study of adaptation using seed set or seedling recruitment as measures of fitness. However, evolutionary importance is not synonymous with ecological importance. The frequency of a trait may change under selection without altering the size of the population (Addicott 1986). The distinction between ecological and evolutionary importance has seldom been recognized in studies of mutualism but is central to predicting extinctions. Pollinator or disperser failure will only affect extinction when populations are seed-limited.

An important measure of relative dependence on seeds is the number of generations needed to persist for some given timespan, say 200–300 years. Traits associated with low risk then include:

1. *Clonal or vegetative propagation*. Clonal plants may live for several thousand years (Richards 1986) and occupy large areas.
2. *Long lifespans*. Plant lifespans vary from weeks to many thousands of years (Loehle 1988). There are several examples of plant populations persisting long after they have lost the capacity to reproduce. *Tilia cordata* reached its northernmost distribution in Britain about 5000 years ago. Northern areas are now too cold for sexual reproduction and recruitment but the populations persist because the trees are 'essentially immortal. . .the stems live for several centuries but, when they collapse, are replaced by vigorous vegetative sprouts' (Pigott 1993). *Ginkgo* may have survived into the modern era because of similar traits (Tredici 1992);
3. *The capacity to resprout vegetatively* after disturbance such as fire or hurricane damage. Plants that persist vegetatively through disturbance are much less dependent on seeds.

Plants with shorter generation times may still have low dependence on seeds if seedling densities greatly exceed the space available for adults (Harper 1977; Anderson 1989; Crawley 1990, 1992). Thus, seed predators are usually poor agents for biological control of weeds because high levels of seed predation achieve little more than the removal of suppressed individuals (Wilson 1964; Harper 1977; Hoffman and Moran 1991). Some populations, however, are strongly influenced by seed predators. Examples include annual herbs (Borchert and Jain 1978; Anderson 1989), shrubs (Louda 1982), especially fire-prone non-sprouting species (Bond 1984; Bond and Slingsby 1984; Cowling *et al.* 1987), mangroves (Smith 1987), and fugitive species from ephemeral habitats (Louda 1989).

Demographic dependence on seeds can be ranked on the basis of these considerations (Table 9.1). The least dependent are those species that, once established, persist indefinitely by clonal propagation or vegetative sprouting. This is one of the easiest traits to identify in the field or from herbarium material or taxonomic monographs. The most seed-dependent species have sparse populations producing few seeds that are short-lived, lack seed banks, and are killed by disturbance.

Compensatory effects

These three factors: the specificity and degree of reproductive dependence on mutualism, and the demographic dependence on seeds, may compensate to reduce the risk of extinction. A study of spring wildflowers in temperate deciduous forests serves as an example (Motten 1986). This community should be highly susceptible to pollinator failure because of the high diversity of insect-pollinated species, the short blooming season before canopy closure, and poor weather that interrupts pollinator activity. This is not the case, as shown by plotting species location on axes of dependence on pollinators against demographic dependence on seeds (Fig. 9.2). Species with the highest demographic dependence on seeds cluster at the low-risk end of the pollinator-dependence axis—they are all self-pollinated or have diverse bee and fly pollinators. At the opposite extreme, species with high-pollinator dependence cluster at the low-risk end of the seed-dependence axis—they are clonal and very long-lived. No species occupies the most vulnerable area shown in the top right corner.

Similar patterns occur in Welsh populations of *Veronica* species (Boutin and Harper 1991). The short-lived seed-dependent species are self-pollinated or self-compatible, while potentially pollinator-limited species are clonal and long-lived (Table 9.2). Extreme examples of pollinator-limited seed set occur in some long-lived clonal species. *Filipendula rubra*, a rare self-incompatible North American herb, has a very low seed set in nature (< 1 viable seed per thousand ovules) but, like other clonal species, genets may be hundreds or even thousands of years old (Aspinwall and Christian 1992).

The frequency of compensatory mechanisms suggests extinction may have already removed high-risk specialists. Alternatively, self-pollination may have evolved rapidly in response to pollinator failure (Motten 1986) or selection may have altered floral morphs to allow visits by alternative pollinators.

Compensation is not universal. A single species of butterfly is the near exclusive pollinator of a guild of 15 spectacular red-flowered species from 4 families (Iridaceae, Orchidaceae, Amaryllidaceae, Crassulaceae) and 7 genera in the fynbos vegetation of South Africa (Johnson and Bond 1992, 1994). Most of the species are incapable of selfing and seed set often fails (Johnson 1992). Demographic dependence on seeds is poorly known for this largely geophytic guild. Most of the streambank species are capable of vegetative propagation. The

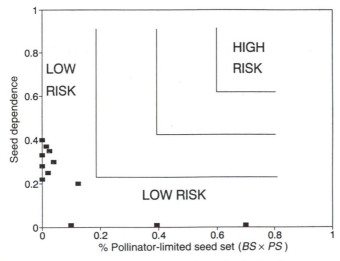

Fig. 9.2 Reproductive compensation in spring flowering herbs of a temperate deciduous forest (Motten 1986). Pollinator-limited seed set, $(BS \times PS)$ $(BS$, breeding system; PS, pollinator specificity) is plotted against demographic dependence on seeds from clonal $= 0$ to short-lived, sparse $= 1$. Contours indicate increasing risk of extinction from the origin.

Table 9.2 Compensatory effects in the genus *Veronica* growing in Britain. Short-lived species have breeding systems or pollination modes which ensure seed set. Self-incompatible species with large, presumably more specialized pollinators are clonal and therefore insured against reproductive failure (Boutin and Harper 1991).

Species	BS	PS	BS × PS	SD	BS × PS × SD
arvensis	0.1	–	<0.1	Annual (1)	c.0
hederifolia	0	–	0	Annual (1)	0
persica	0.1	–	<0.1	Annual (1)	c.0
serpyllifolia	0.5	0.2	0.1	Clonal (0)	0
chamaedrys	1	0.5	0.5	Clonal (0)	0
filiformis	1	0.5	0.5	Clonal (0)	0

BS, breeding system; PS, pollinator specificity; SD, seed dependence; $BS \times PS \times SD$, extinction risk

bulbs of amaryllids are probably longer lived (e.g., Snijman 1992) than irid corms (J. Vlok unpublished). The orchids produce hundreds of tiny seeds from a single capsule and may be site-rather than seed-limited (Calvo 1993). Figure 9.3 ordinates the species, on the basis of available information, on axes of reproductive dependence on the butterfly versus demographic dependence on seeds. Unlike the temperate herb flora, nearly half of the butterfly guild has no compensatory mechanism for surviving pollinator failure. This may have caused

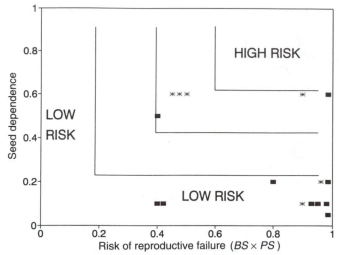

Fig. 9.3 Compensatory effects in a guild of species all pollinated by a single butterfly species, *Meneris tulbaghiae*, in Cape fynbos (see Johnson and Bond 1994). Endangered species are indicated by asterisks. Contours indicate increasing risk of extinction from the origin.

the extreme rarity of a large proportion of the species (Fig. 9.3). Alternatively, the syndrome may have evolved because the butterfly is a reliable pollinator of sparse populations (Johnson 1992).

Compensatory mechanisms in dispersal are much more difficult to identify. They imply that species with obligatory dependence on dispersal persist because they have reliable (generalist?) dispersers or are not seed-limited. Platt and Weiss (1985) studied a group of fugitive species colonizing disturbances created by badgers in the American prairies. They found both a high dependence on dispersal and high demographic dependence on seeds for persistence but all five species in the guild are wind-dispersed.

9.6 A vulnerability index for population consequences of mutualism

The three concerns, risk of pollinator or disperser failure, reproductive dependence on the mutualism, and demographic dependence on seeds, can be combined in a subjective index for rapid assessment of extinction risk. A simple formula which caters for compensatory effects by taking products is

$$VI = [(BS \times PS) + (DD \times DS)] \times [SD]$$

Here, *VI* is a vulnerability index, *BS* is the breeding system, *PS* for pollinator specificity, *DD* for disperser dependence, *DS* for disperser specificity, and *SD* (seed dependence) for demographic dependence on seeds. The first term in brackets on the right-hand side refers to reproductive effects, and the second to demographic effects. Species are ranked for each class using the criteria listed in Table 9.1. For example, a self-pollinated wind dispersed annual has a $VI = [0 \times y + z \times 0] \times [1] = 0$, where y and z may be unknown but there is obviously no population consequence of failed mutualism. In contrast, a rare, butterfly-pollinated, short-lived *Gladiolus* may have a $VI = [1 \times 1 + y \times 0] \times [1] = > = 1$, and the species is flagged as vulnerable. Note that partial information is adequate and that any value approaching 1 implies vulnerability. The index is useful both as a heuristic tool for considering the population consequences of mutualism or as a practical way of quickly assessing vulnerable species.

It is possible to use either pollination or dispersal information or both depending on completeness of information. Breeding systems and pollinator and disperser specificity can be derived from field information or herbaria records when these are not available using the criteria listed in Table 9.1. Consideration of seed dependence is essential if there is concern for imminent extinction. Some indication of dependence on seeds can be gleaned from herbaria or systematic monographs, by scoring for traits, such as sprouting capacity, clonal propagation or seed size, but field observations of criteria listed in Table 9.1 would be more valuable.

9.7 Case studies

The risk of extinctions due to collapse of mutualisms varies greatly among taxa and region. In highly seasonal climates, most plants seem well insured against pollinator failures. Wind pollination, self-pollination, and asexual propagation are common at high latitudes and altitudes (Richards 1986; Regal 1982). Entomophilous flowers in temperate areas often have very diverse pollinator faunas and are well buffered against pollinator failure (Schemske 1983; Herrera 1988; Howe and Westerly 1988). The compensatory effects illustrated in Fig. 9.2 and Table 9.2 may therefore be quite common.

Cape Proteaceae

An interesting contrast is seen in the fire-prone shrublands of Australia and South Africa. Proteaceae are prominent in the extremely species-rich fynbos flora of the Cape. According to the 'vulnerability' index, nearly 50% of the more than 300 species in the Cape flora are threatened with extinction should their mutualisms collapse. This is because the populations of many species appear to be strongly seed-limited: most do not sprout after fire, seeds are few and large, and seedling populations are sparse with little evidence for self-thinning (Le Maitre and Midgley 1992). Despite this risky lifestyle, most are well buffered against pollination failure. The dioecious genera are either wind-pollinated or pollinated by numerous insects, especially beetles (Hattingh and Giliomee 1989). Of the hermaphrodite genera, many species of *Protea*, *Mimetes*, and *Leucospermum* have large showy inflorescences pollinated by birds. However, exclusion experiments have shown that *Protea* species are also pollinated effectively by a variety of insects, especially beetles (Coetzee and Giliomee 1985; see also Vaughton 1992 for Australian *Banksia*).

Seed dispersal in fynbos Proteaceae is either by wind or ants. The experimental studies suggest that the ant-dispersed species have a near obligate dependence on dispersal to escape rodent predation or fire (Bond and Slingsby 1984; Slingsby and Bond 1985). At least one ant-dispersed species behaves as a fugitive since parental sites are invaded by more competitive wind-dispersed species after fire (Yeaton and Bond 1991). A variety of ant species disperse seeds suggesting a low risk of dispersal failure. However, small parts of the region have been invaded by the Argentine ant, *Iridomyrmex humilis*, which displaces most native species. It does not disperse seeds into ant nests with the consequence of heavy seed losses and very poor seedling recruitment in invaded areas (Bond and Slingsby 1984). Should *Iridomyrmex* continue to spread, nearly half the fynbos Proteaceae may be lost.

This example clearly illustrates the dilemma in estimating extinction risk based only on dispersal specificity. One never knows which group of mutualists will be threatened nor how complete local extinction may be. In this case, the near obligate dependence on the mutualism and the demographic dependence on seeds are better indicators of vulnerability.

Fynbos is also rich in extraordinary and highly specific pollination mutualisms. Many of the most beautiful plants have narrowly specific associations with a sparse and unusual pollinator fauna, including long-tongued flies, oil-collecting bees, monkey beetles, large carpenter bees or the butterfly *Meneris tulbaghiae* (Rebelo 1987; Johnson 1992; Steiner 1989; Manning and Linder 1992). These species seem particularly vulnerable to pollinator failure. However, information on pollinators, breeding systems, and demographic dependence on seeds is too limited for any general assessment at this stage. In a recent analysis of threats facing the flora, not a single Red Data Book species was listed as threatened by extinction of its mutualist partner (Rebelo 1992). This probably says more about the cryptic nature of extinctions caused by failed mutualism than indicating the true magnitude of the problem.

Tropical rainforests

Tropical forests may also be high-risk floras. Lowland tropical forests in particular are marked by unusually low levels of self-pollination and very high levels of dioecy (Table 9.3; Bawa *et al.* 1985*a*). In contrast to temperate regions, lowland tropical forests have much more specialized pollinator and disperser relationships (e.g., Regal 1982). Obligate one to one pollinator relationships (such as figs and fig wasps) are the exception in the tropics but the majority of species are pollinated by only one or a few species belonging to the same taxonomic group (e.g., euglossines, humming-birds, scarab beetles, bats, etc.) (Bawa 1990). The combination of self-incompatibility and pollinator specificity indicates a high degree of reproductive dependence on mutualism.

The same may apply for dispersal. Neotropical rainforests have a very high

Table 9.3 Distribution of breeding systems (% of tree species) in tropical forest trees (ex Bawa 1990). Tropical lowland rain forests (TLRF) have the highest dependence on pollinators for seed set

Forest type	Self-comp.	Self-incompat.	Dioecious
TLRF, Costa Rica	20	80	23
Montane forest (Venezuela)	62	28	31
Montane forest (Jamaica)	85	15	21

Table 9.4 Comparison of fruit types (% of species) in tropical and deciduous forest canopy trees with sub-canopy trees in parentheses. Lowland neotropical forests and sub-canopy trees have a high proportion of animal-dispersed seeds. (Data from Howe and Smallwood 1982.)

Forest type	Fleshy		Abiotic		Nuts	Other
Neotropics (aseasonal)	89		3			8
Neotropics (wet)	89	(94.5)	10	(5.5)		1
Neotropics (montane)	79	(87)	21	(13)		
Neotropics (dry)	57.5	(77)	33.5	(23)		9
Palaeotropics (Asia, Africa)	41.5	(80)	36.5	(7)		22 (13)
Temperate deciduous	19	(92.5)	35		46	0 (7.5)

proportion of fleshy fruits in canopy trees (Howe and Westerley 1988), less so in Asian and African forests, but all forests have a high incidence of fleshy fruits in the sub-canopy (Table 9.4). At least some studies suggest near obligate dependence on dispersal for recruitment so that reproductive dependence on mutualisms may also be high for dispersal (Howe and Westerley 1988; Augspurger 1984).

There is much less information on demographic dependence on seeds. Many tree populations are sparse with little density-dependent mortality (Hubbel and Foster 1990). Fugitive species (*Cecropia*) exploiting light gaps may be seed-limited and therefore particularly vulnerable to disruption of pollinators or dispersers (Garwood 1989). Tropical trees are difficult to age and I have no information on the incidence of sprouting after disturbance. If sprouting is rare, then the cascade of extinctions envisaged by Janzen (1974, 1987) may be true for plants as well as their animal partners.

9.8   Examples of pollinator and disperser extinction

There are several convincing cases of complete extinction of animal partners in reproductive mutualisms but very few of subsequent plant extinctions. Janzen has argued persuasively for extinction of a guild of mammals that dispersed large fruits in the neotropics and arid lands of North America but the plant species still persist thousands of years after the loss of their dispersers (Janzen and Martin 1982; Janzen 1986). The oil-collecting bee pollinator of a rare fynbos shrub, *Ixianthes*, has become locally extinct but the plant sprouts and is not immediately threatened with extinction (Steiner 1993). Temple (1977) argued that recruitment of a Mauritian tree may have failed because its seeds needed processing by the extinct dodo. However, the tree still survives on Mauritius. In Hawai'i, extinction of native bird pollinators resulted in a change of pollinators for the ieie, *Freycinetia arborea*, not extinction (Cox 1983). It is arguable whether the persistence of plant species for hundreds, if not thousands of years, after extinction of their mutualist partners is evidence for general resilience to extinction or just the tip of the iceberg of all the other species that went extinct.

9.9   Conclusions

This analysis of the importance of an ecological interaction for predicting extinction differs from previous ecological studies of extinction risk. It cannot make broad generalizations, such as those offered by island biogeographic theory, nor explicit probabilities of extinction time, such as population vulnerability analysis. However, it does identify species threatened by the extinction of interactions. The approach explicitly takes into account compensatory effects that may reduce the threat. Taxonomic databases can be useful for partial assessments of vulnerability but should be supplemented by field surveys. The point is that the approach is feasible—it can be included in surveys of extinction threats. Such surveys would be useful, too, in identifying key animal partners whose demise might trigger a spate of plant extinctions and which therefore warrant special attention.

The longer view

I have taken a short-term view of the impact of animal extinctions on plant extinctions. The importance of interactions for species persistence has been ignored for too long. However, lists of threatened plant species are not the true measure of the problem for conservation. Few would disagree that the extinction of the interaction itself is of much greater concern. It will be tragic if the remaining natural areas of the world are filled with ageing plants silent as graveyards with no butterfly or sunbird pollinators working their flowers or large colourful birds eating their fruits.

Summary

There is an extensive literature on pollination and dispersal, very little of which deals with the consequences of reproductive failure or its most extreme consequence—extinction. The risk of plant extinctions can be assessed by considering the probability of dispersal or pollinator failure, the dependence of reproduction on mutualisms, and the dependence of demography on seeds. Traits for rapidly ranking species according to these three criteria are indicated. Analysis of case studies suggests that plants often compensate for high risk in one of the three categories by low risk in another. For example, self-incompatible plants with rare specialist pollinators often propagate vegetatively. Some systems, including elements of the Cape flora and lowland tropical rainforest, lack compensatory traits and the risk of plant extinction from failed mutualism is high. The vulnerability assessment can be used to identify both plant groups at risk and key mutualist partners.

References

Adams, P. B. & Lawson, S. D. (1993). Pollination in Australian orchids: a critical assessment of the literature 1882–1992. *Aust. J. Bot.*, **41**, 553–76.

Addicott, J. F. (1986). On the population consequences of mutualism. In *Community ecology*, (ed. J. Diamond and T. Case), pp. 425–36. Harper & Row, New York.

Anderson, A. N. (1989). How important is seed predation to recruitment in stable populations of long-lived perennials? *Oecologia*, **81**, 310–15.

Aspinwall, N. and Christian, T. (1992). Pollination biology, seed production, and population structure in Queen of the Prairie, *Filipendula rubra* (Rosaceae) at Botkin Fen, Missouri. *Amer. J. Bot.*, **79**, 488–94.

Augspurger, C. K. (1984). Seedling survival among tropical tree species: Interactions of dispersal distance, light gaps and pathogens. *Ecology*, **65**, 1705–12.

Bawa, K. S. (1990). Plant–pollinator interactions in tropical rain forests. *Ann. Rev. Ecol. Syst.*, **21**, 399–422.

Bawa, K. S., Perry, D. R., and Beach, J. H. (1985*a*). Reproductive biology of tropical lowland rain forest trees. I. Sexual systems and incompatibility mechanisms. *Amer. J. Bot.*, **72**, 331–45.

Bawa, K. S., Bullock, S. H., Perry, D. R., Coville, R. E., and Grayum, M. H. (1985*b*). Reproductive biology of tropical lowland rain forest trees. II. Pollination systems. *Amer. J. Bot.*, **72**, 346–56.

Beattie, A. J. (1985). *The evolutionary ecology of ant–plant mutualisms*. Cambridge University Press.

Bond, W. J. (1984). Fire survival of Cape Proteaceae—influence of fire season and seed predators. *Vegetatio*, **56**, 65–74.

Bond, W. J. and Slingsby, P. (1984). Collapse of an ant–plant mutualism: The Argentine ant, *Iridomyrmex humilis* and myrmecochorous Proteaceae. *Ecology*, **65**, 1031–7.

Borchert, M. I. and Jain, S. K. (1978). The effect of rodent seed predation on four species of California annual grasses. *Oecologia*, **33**, 101–13.

Boucher, D. H. (ed.) (1985). *The biology of mutualism: ecology and evolution*. Oxford University Press.

Boutin, C. and Harper, J. L. (1991). A comparative study of the population dynamics of five species of *Veronica* in natural habitats. *J. Ecol.*, **79**, 199–221.

Breytenbach, G. J. (1986). Impacts of alien organisms on terrestrial communities with emphasis on communities of the south-western Cape. In *The ecology and management of biological invasions in southern Africa*, (ed. I. A. W. Macdonald, F. J. Kruger, and A. A. Ferrar), pp. 229–38. Oxford University Press.

Calvo, R. N. (1993). Evolutionary demography of orchids: intensity and frequency of pollination and the cost of fruiting. *Ecology*, **74**, 1033–42.

Coetzee, J. H. and Giliomee, J. H. (1985). Insects in association with the inflorescences of *Protea repens* (L.) (Proteaceae) and their role in pollination. *J. Entomol. Soc. S. Afr.*, **48**, 303–14.

Cowling, R. M., Lamont, B. B., and Pierce, S. M. (1987). Seed bank dynamics of four co-occurring *Banksia* species. *J. Ecol.*, **75**, 289–302.

Cox, P. A. (1983). Extinction of the Hawaiian avifauna resulted in a change of pollinators for the ieie, *Freycinetia arborea*. *Oikos*, **41**, 195–9.

Crawley, M. J. (1990). Plant population dynamics. *Phil. Trans. Roy. Soc. Lond.*, **B314**, 125–40.

Crawley, M. J. (1992). Seed predators and plant population dynamics. In *Seeds: the ecology of regeneration in plant communities*, (ed. M. Fenner), pp. 105–56. CAB International, Wallingford, UK.

Cruden, R. W. (1977). Pollen–ovule ratios: a conservative indicator of breeding systems in flowering plants. *Evolution*, **31**, 32–46.

Dafni, A. (1992). *Pollination ecology: a practical approach*. Oxford University Press.

Dafni, A. *et al.* (1990). Red bowl-shaped flowers: convergence for beetle pollination in the Mediterranean region. *Israel. J. Bot.*, **39**, 81–92.

Diamond, J. M. (1984). 'Normal' extinctions of isolated populations. In *Extinctions*, (ed. M. H. Nitecki), pp. 191–246. University of Chicago Press.

Estrada, A. and Fleming, T. H. (ed.) (1986). *Frugivores and seed dispersal*. Junk, Dordrecht.

Faegri, K., and van der Pijl, L. (1979). *The principles of pollination ecology*. Pergamon, Oxford.

Feinsinger, P. (1983). Coevolution and pollination. In *Coevolution*, (ed. D. J. Futuyma and M. Slatkin), pp. 282–310. Sinauer, Sunderland, MA.

Feinsinger, P. (1987). Approaches to nectarivore-plant interactions in the New World. *Rev. Chil. Hist. Nat.*, **60**, 285–319.

Garwood, N. C. (1989). Tropical seed banks: a review. In *Ecology of seed banks*, (ed. M. A. Leck, V. T. Parker, and R. L. Simpson), pp. 149–210. Academic Press, New York.

Gess, F. W. and Gess, S. K. (1993). Effects of increasing land utilization on species representation and diversity of aculeate wasps and bees in the semi-arid areas of southern Africa. In *Hymenoptera and biodiversity*, (ed. J. LaSalle and I. D. Gauld), pp 83–113. CAB International, Wallingford, UK.

Gilbert, L. E. (1980). Food web organization and the conservation of neotropical diversity. In *Conservation biology*, (ed. M. E. Soule and B. A. Wilcox), pp. 11–33. Sinauer, Sunderland, MA.

Ha, C. O., Sands, V. E., Soepadmo, E., and Jong, K. (1988). Reproductive patterns of selected understorey trees in the Malaysian rain forest: the apomictic species. *Bot. J. Linn. Soc.*, **97**, 317–31.

Harper, J. L. (1977). *Population biology of plants*. Academic Press, London.

Hattingh, V. and Giliomee, J. H. (1989). Pollination of certain *Leucadendron* species (Proteaceae). *S. African J. Bot.*, **55**, 387–93.

Herrera, C. M. (1984). A study of avian frugivores, bird-dispersed plants, and their interaction in Mediterranean scrublands. *Ecol. Monogr.*, **54**, 1–23.

Herrera, C. M. (1988). Variation in mutualisms: the spatiotemporal mosaic of a pollinator assemblage. *Biol. J. Linn. Soc.*, **35**, 95–125.

Hoffmann, J. H. and Moran, V. C. (1991). Biocontrol of a perennial legume, *Sesbania punicea*, using a florivorous weevil, *Trichaphion lativentre*: weed population dynamics with a scarcity of seeds. *Oecologia*, **88**, 574–6.

Howe, H. F. (1984). Implications of seed dispersal by animals for the management of tropical reserves. *Biol. Conserv.*, **30**, 261–81.

Howe, H. F. and Smallwood, J. (1982). Ecology of seed dispersal. *Ann. Rev. Ecol. Syst.*, **13**, 201–28.

Howe, H. F. and L. C. Westerley (1988). *Ecological relationships of plants and animals.* Oxford University Press.

Hubbel, S. P. and Foster, R. B. (1986). Biology, chance and history of the structure of tropical rain forest tree communities. In *Community ecology*, (ed. J. Diamond and T. Case), pp. 314–30. Harper and Row, New York.

Hughes, L. and Westoby, M. (1992). Effect of diaspore characteristics on removal of seeds adapted for dispersal by ants. *Ecology*, **73**, 1300–12.

Janzen, D. H. (1974). The deflowering of Central America. *Nat. Hist.*, 49–53.

Janzen, D. H. (1983). Dispersal of seeds by vertebrate guts. In *Coevolution*, (ed. D. J. Futuyma and M. Slatkin), pp. 232–62. Sinauer, Sunderland, MA.

Janzen, D. H. (1986). Chihuahuan desert Nopaleras: defaunated big mammal vegetation. *Ann. Rev. Ecol. Syst.*, **17**, 595–636.

Janzen, D. H. (1987). Insect diversity of a Costa Rican dry forest: why keep it and how? *Biol. J. Linn. Soc.*, **30**, 343–56.

Janzen, D. H. and Martin, S. (1982). Neotropical anachronisms: the fruits the gomphotheres ate. *Science*, **215**, 19–27.

Jennersten, O. (1988). Pollination of *Dianthus deltoides* (Caryophyllaceae): effects of habitat fragmentation on visitation and seed set. *Conserv. Biol.*, **2**, 359–66.

Johansen, C. A. (1977). Pesticides and pollinators. *Ann. Rev. Entomol.*, **22**, 177–92.

Johnson, S. D. (1992). Plant–animal relationships. In *The ecology of fynbos*, (ed. R. M. Cowling), pp. 175–205, Oxford University Press.

Johnson, S. D. and Bond, W. J. (1992). Habitat dependent pollination success in a Cape orchid. *Oecologia*, **91**, 455–6.

Johnson, S. D. and Bond, W. J. (1994). Red flowers and butterfly pollination in the fynbos. In *Plant animal interactions in Mediterranean-type ecosystems*, (ed. M. Arianoutsou and R. H. Groves), pp. 51–68. Kluwer, Dordrecht.

Jordano, P. (1992). Fruits and frugivory. In *Seeds: the ecology of regeneration in plant communities*, (ed. M. Fenner), pp. 105–56. CAB International, Wallingford, UK.

Kevan, P. G. (1975). Pollination and environmental conservation. *Environ. Conserv.*, **2**, 293–8.

Kevan, P. G., Thomson, J. D., and Plowright, R. C. (1985). Matacil insecticide spraying, pollinator mortality, and plant fecundity in New Brunswick forests. *Can. J. Bot.*, **63**, 2056–61.

Le Maitre, D. C. and Midgley, J. J. (1992). Plant reproductive ecology. In *The ecology of fynbos: nutrients, fire and diversity*, (ed. R. M. Cowling), pp. 135–74. Oxford University Press.

Linhart, Y. B. and Feinsinger, P. (1980). Plant–hummingbird interactions: effects of island size and degree of specialization on pollination. *J. Ecol.*, **68**, 745–60.

Loehle, C. (1988). Tree life history strategies: the role of defenses. *Can. J. For.*, **18**, 209–22.

Louda, S. M. (1982). Distribution ecology: Variation in plant recruitment over a gradient in relation to insect seed predation. *Ecol. Monogr.*, **52**, 25–41.

Louda, S. M. (1989). Predation in the dynamics of seed regeneration. In *Ecology of seed banks*, (ed. M. A. Leck, V. T. Parker, and R. L. Simpson), pp. 25–51. Academic Press, New York.

Manning, J. C. and Linder, H. P. (1992). Pollinators and evolution in *Disperis* (Orchidaceae), or why are there so many species? *S. African J. Sci.*, **88**, 38–49.

Martin, T. E. (1985). Resource selection by tropical frugivorous birds: Integrating multiple interactions. *Oecologia*, **66**, 563–73.

Motten, A. F. (1986). Pollination ecology of the spring wildflower community of a temperate deciduous forest. *Ecol. Monogr.*, **56**, 21–42.

Pierce, S. M. and Cowling, R. M. (1991). Dynamics of soil-stored seed banks of six shrubs in fire-prone dune fynbos. *J. Ecol.*, **79**, 731–47.

Pigott, C. D. (1993). Are the distributions of species determined by failure to set seed? In *Fruit and seed production*, (ed. C. Marshall and J. Grace), pp. 203–16. Cambridge University Press.

Platt, W. J. and Weis, I. M. (1985). An experimental study of competition among fugitive prairie plants. *Ecology*, **66**, 708–20.

Plitmann, U. and Levin, D. A. (1990). Breeding systems in the Polemoniaceae. *Pl. Syst. Evol.*, **170**, 205–14.

Price, M. V. and Jenkins, S. H. (1986). Rodents as seed consumers and dispersers. In *Seed dispersal*, (ed. D. R. Murray), pp. 191–235. Academic Press, Australia.

Rebelo, A. G. (ed.) (1987). *A preliminary synthesis of pollination biology in the Cape flora*, SANSP Report, No. 141. CSIR, Pretoria.

Rebelo, A. G. (1992). Red Data Book species in the Cape Floristic Region: threats, priorities and target species. *Trans. Roy. Soc. S. Afr.*, **48**, 55–86.

Regal, P. J. (1982). Pollination by wind and animals: ecology of geographic patterns. *Ann. Rev. Ecol. Syst.*, **13**, 497–524.

Richards, A. J. (1986). *Plant breeding systems*. Allen and Unwin, London.

Schemske, D. W. (1983). Limits to specialization and coevolution in plant–animal mutualisms. In *Coevolution*, (ed. M. Nitecki), pp. 67–109. University of Chicago Press.

Slingsby, P. and Bond, W. J. (1985). The influence of ants on the dispersal distance and seedling recruitment of *Leucospermum conocarpodendron* (Proteaceae). *S. African J. Bot.*, **51**, 30–4.

Smith, T. J. (1987). Seed predation in relation to tree dominance and distribution in mangrove forests. *Ecology*, **68**, 266–73.

Snijman, D. A. (1992). Systematic studies in the tribe Amaryllideae (Amaryllidaceae). Unpublished thesis, University of Cape Town.

Steiner, K. E. (1989). The pollination of *Disperis* (Orchidaceae) by oil collecting bees in southern Africa. *Lindleyana*, **4**, 164–83.

Steiner, K. E. (1993). Has *Ixianthes* (Scrophulariaceae) lost its special bee? *Pl. Syst. Evol.*, **185**, 7–16.

Temple, S. A. (1977). Plant animal mutualism: coevolution with dodo leads to near extinction of plant. *Science*, **197**, 885–6.

Tredici, P. D. (1992). Natural regeneration of *Ginkgo biloba* from downward growing cotyledonary buds (basal chichi). *Amer. J. Bot.*, **79**, 522–30.

Vaughton, G. (1992). Effectiveness of nectarivorous birds and honeybees as pollinators of *Banksia spinulosa* (Proteaceae). *Aust. J. Ecol.*, **17**, 43–50.

Vogel, S. and Westerkamp, C. (1991). Pollination: an integrating factor of biocenoses. In *Species conservation: a population-biological approach*, (ed. A. Seitz and V. Loeschcke), pp. 159–70. Birkhauser, Basel.

Wheelwright, N. T. (1985). Fruit size, gape width, and the diets of fruit eating birds. *Ecology*, **66**, 808–18.

Wilson, F. (1964). The biological control of weeds. *Ann. Rev. Entomol.*, **9**, 225–44.

Yeaton, R. I. and Bond, W. J. (1991). Competition between two shrub species: dispersal differences and fire promote coexistence. *Amer. Nat.*, **138**, 328–41.

10

Population dynamic principles

John H. Lawton

To admit that species generally become rare before they become extinct, to feel no surprise at the rarity of the species, and yet to marvel greatly when the species ceases to exist, is much the same as to admit that sickness in the individual is the forerunner of death—to feel no surprise at sickness, but, when the sick man dies, to wonder and to suspect that he died of some deed of violence.

(Charles Darwin, The origin of species, Chapter 11)

Who killed Cock Robin?
I, said the Sparrow,
With my bow and arrow,
I killed Cock Robin.

(Traditional English Nursery Rhyme)

10.1 Introduction

The death of an individual, a population, or an entire species can have many causes, even if one ultimately delivers the *coup de grâce*. Are there, however, general rules and constraints that may help us better understand rarity and extinction?

Simberloff (1986), Cracraft (1992), and Caughley (1994) all provide succinct reviews of the literature on extinctions, and seek to draw some general conclusions. It is useful, for example, despite the opening quotation from Darwin, to draw a distinction between processes that make populations rare in the first place (what Simberloff calls the 'ultimate causes of extinction', and Caughley refers to as 'the declining population paradigm'), and processes that may finally cause extinction, once populations are small (Simberloff's 'proximate causes', and Caughley's 'small population paradigm'). Proximate causes are why small populations still die out, even when protected. They are generally agreed to include demographic and environmental stochasticity, genetic deterioration, and social dysfunction, although the relative contributions of these four harbingers of doom are not well understood.

Ultimate causes of extinction—why species are, or become rare in the first place—are a different matter, although one thing is depressingly clear. Throughout human history, formerly widespread and abundant species have been made rare and vulnerable to the proximate causes of extinction by hunting, habitat destruction, and pollution. On the other hand, some species are naturally rare (many island endemics, for example), although they too may be made even rarer by man. It is currently unclear whether naturally rare, and anthropogenically rare species differ in fundamental ways to the risks of extinction from proximate causes. I touch on this problem at the end of the chapter.

It is not my intention to repeat existing reviews of the four proximate causes of extinction. Instead, I will review a number of less familiar, and sometimes contentious issues that may contribute to both ultimate and proximate causes. I am particularly concerned to see if there are general patterns and rules that may make particular taxa, in particular places, more or less prone to rarity and ultimately to extinction. I have not, generally, made a distinction between the local extinction of populations, and the global extinction of species; as Andrewartha and Birch pointed out many years ago (1954, p. 665): 'there is no fundamental distinction to be made between the extinction of a local population and the extinction of a species other than. . .the species becomes extinct with the extinction of the last local population.'

10.2   Population abundance and geographic range

Local abundance and size of range

Rarity, which predisposes populations or entire species to extinction via proximate causes, is actually a rather complicated phenomenon; there are several ways of being rare (Rabinowitz et al. 1986), involving both the size of a species' geographic range, and population abundance within that range. Interestingly, on average, population abundances and species' geographic ranges are not independent entities (Brown 1984; Hanski 1982; Lawton 1993). Within particular taxa and geographic regions, species with large ranges tend to have greater local abundances at sites where they occur than do geographically more restricted species. Examples include plants, birds, mammals, fish, and a variety of invertebrates from molluscs and mites to zooplankton and insects (Gaston and Lawton 1990). There is usually considerable unexplained variation in these plots, so that an individual species can be widespread but rare everywhere, or locally common but with a small total geographic range (Rabinowitz et al. 1986). Nevertheless, the general pattern appears to be robust, despite problems with definitions of geographic range (Gaston 1991), concerns about sampling artefacts, and occasional negative correlations (Gaston and Lawton 1990) that occur under somewhat unusual circumstances (see Lawton 1993 for a brief discussion of these problems).

Despite its theoretical and practical interest, the positive correlation between range and abundance is not well understood. Theoretically, there are a number of ways in which such a correlation might be generated (Lawton 1993; Hanski et al. 1994). The simplest, proposed by Brown (1984), is that species able to exploit a wide range of resources (species with 'broad niches') become both widespread and locally abundant. Several metapopulation dynamic models also predict positive correlations between geographic range measured as the number (or proportion) of patches occupied and population density within patches (Gyllenberg and Hanski 1992). Positive correlations in these models are a product of the 'rescue effect' (immigration reduces the risk of local extinction), mortality during migration, and difficulties in establishing new populations.

A broadly positive correlation between size of range and population abundance carries two important messages when assessing extinction risks. First, and other things being equal, species that are most at risk from proximate causes are those with small geographic ranges, because they will, on average, also be locally rare, even in areas where they occur. Nature has double jeopardy built into her game plan.

Second, this double jeopardy may be particularly serious when populations and ranges are artificially reduced and threatened by ultimate causes of extinction. Reductions in range, via habitat destruction or extirpation by hunting, ought, eventually to result in a reduction in population density in surviving populations, even when protected. And reversing these arguments, overall reductions in population density (e.g., by pollution, hunting, or for migrants, problems on the wintering grounds) should result in correlated reductions in range size, in the absence of any habitat destruction. These empirical predictions are confirmed by metapopulation dynamic models (e.g., Lande 1991; Gyllenberg and Hanski 1992; Lawton *et al.* 1994) but they are not easily derived from Brown's explanation for the positive correlation between range and abundance, and I know of no unequivocal experimental tests. I find it remarkable that such an important problem has not attracted more experimental and practical attention.

Tropical versus temperate ranges and abundances

A second empirical pattern that may influence extinction risk is an apparent decline in the average sizes of species' ranges within comparable taxa, as one moves from higher to lower latitudes (Rapoport 1982). Stevens (1989) and France (1992) document declines in size of range towards the tropics for groups as diverse as trees, molluscs, crustacea, fish, amphibians and reptiles, and mammals [although not all recent analyses confirm the pattern (Ricklefs and Latham 1992); see Lawton *et al.* (1994) for a brief review].

A decline in average range sizes from the poles to the tropics, need not, of course, mean that average local population abundances of comparable taxa are also lower in the tropics; too many other key variables change along this same gradient. It is therefore intriguing that latitude alone explains 46.7% of variation in population densities in a major compendium of data assembled by Currie and Fritz (1993). Average population densities increase linearly from the Equator towards the poles in invertebrates, ectothermic vertebrates, mammals, and birds. At any one latitude, densities range over four or more orders of magnitude within these groups, but the overall trend for populations, on average, to have lower densities at lower latitudes is clear, and consistent with a decline in range sizes towards the tropics.

Taken at face value, both trends must make tropical taxa more prone to extinction from human activities, particularly habitat destruction, compared with equivalent temperate taxa. Starkly, clearing 1000 km^2 of tropical forest will, on average, pose a much greater threat of extinction to inhabitants living at low average densities, in small ranges, than clearing an equivalent area of temperate forest; of the 1029 species of birds in the world listed in Rands (1991) as threatened with global extinction, 442 live in tropical forests—more than twice as many as in the next most important habitat, wetlands (although many other factors also contribute to this depressing statistic).

10.3   Variations in abundance within geographic ranges

Textures of abundance

One reason why the correlation between size of geographic range and local abundance is fuzzy is that species are not evenly distributed throughout their range (Brown 1984; Wiens 1989). Very crudely, densities are often envisaged as being greatest near the centre of a range, and least near the boundaries (e.g., Hengeveld and Haeck 1982). In practice, textures of distribution and abundance are often more complex than a gradual decline from the centre to the edge of the range, with multimodal patterns of abundance being common and perhaps even the norm (Taylor and Taylor 1979; Brown 1984; Root 1988; Wiens 1989; Margules and Austin, Chapter 12).

Because average abundances vary across species' ranges, it follows that one or more of the key demographic rates (birth, death, immigration, and emigration) also change across the range, in response to changes in environment and resources (Richards 1961; Huffaker and Messanger 1964). Some recently documented examples are provided by Randall (1982), Rogers and Randolph (1986), and Caughley et al. (1988). At some point close to the range boundary, rates of population increase from low densities (r) must on average be zero. Beyond the point where r = zero, 'sink populations' (Pulliam 1988) with negative average r may be sustained by immigration from 'source populations' in which overall population performance (but not necessarily density) is higher (Fig. 10.1).

Implications for extinction

One interesting consequence of variations in population processes across a species' range is that attempts to restore populations by re-introductions into historical, but currently unoccupied parts of a range are less likely to fail (the re-introduced population is less likely to become extinct) in the core of the former distribution than on its periphery or beyond it. Consistent with this prediction, 76% of 133 documented translocations (re-introductions) of birds and mammals into former core areas succeeded, compared with 48% of 54 translocations to the periphery or beyond (Griffith et al. 1989).

It also follows that wholesale persecution or habitat destruction may leave isolated populations of high conservation importance in marginal habitats (the

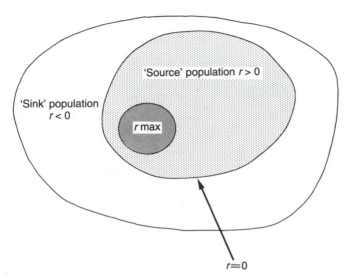

Fig. 10.1 Idealized geographic range. The species is assumed to have a maximum intrinsic rate of increase, r, towards the centre of its range (NB, it does not follow that this is also the region with maximum abundance). Values decline away from this central region, to a point where average rates of increase are zero; populations and individuals may occur beyond this region (in the unshaded zone), but only as 'sinks' from 'source' populations deeper within the (shaded) range (Lawton 1993).

'where we find them now is not where they want to be' phenomenon) (Fig. 10.2). Relict populations of bontebok (*Damaliscus dorcas* subsp. *dorcas*), an antelope, in the fynbos of Cape Province, South Africa (van Wilgen *et al.* 1992) and of red kites (*Milvus milvus*), a raptor, in south-central Wales on the very edge of its European range (Cramp and Simmons 1980) are good examples. Such populations may have very low rates of increase (well below that which the species can achieve in better habitats), making their survival even more precarious. *In extremis* some populations may persist only because of immigration from the core; destruction of the source guarantees the ultimate extinction of the sink population, making its conservation impossible (Harrison 1991; Harrison *et al.* 1988).

Finally, as noted above, if overall population numbers decline because of falling birth rates or rising death rates, we expect ranges to contract anyway, even in the absence of habitat destruction. If the original range had a single, well-defined centre, ranges should collapse towards that core; if there were originally multiple modes, contraction and fragmentation into former 'hotspots' seems more likely (Fig. 10.2).

For reasons that are not entirely clear, there have been rather few attempts to document and link patterns of population decline with changes in species' distributions. Hengeveld (1989) shows that declines in European populations of fir trees (*Abies* spp.) were accompanied by range fragmentation. A drastic decline

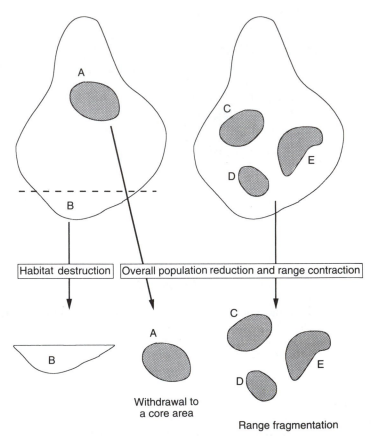

Fig. 10.2 Population abundances vary across species ranges, tending to be higher towards the (shaded) centre of the range; peak abundances may be reached in just one area (A) or several different parts of the range (C–E). Habitat destruction may leave a species surviving only in very marginal habitats (B). Overall reductions in population abundances may cause contraction of species ranges to a single region (A), or to range fragmentation (C–E) (Lawton 1993).

during the 20th century by British populations of a skipper butterfly, *Hesperia comma*, left a scattered and highly fragmented populations in 46 or fewer refugia by the mid 1950s (Thomas and Jones 1993). But as Wilcove and Terborgh (1984) show for some North American birds, range fragmentation does not always accompany decline. The highly endangered Kirtland's warbler (*Dendroica kirtlandii*), withdrew to the historical centre of its range, leaving peripheral areas virtually empty, as populations collapsed by 60% between 1961 and 1971.

Which type of range decline poses the greater threat of ultimate extinction? Fragmentation and isolation may exacerbate population declines in remaining areas via metapopulation processes; contraction towards a single core area may avoid these problems but puts all the conservation eggs into one geographical basket. A growing body of data and theory addresses the problems of extinction risk in a few large (in the limit, one) and many small, fragmented populations, and it is to these questions that we now turn.

Extinction risks in one to few large populations versus many small populations

Imagine two species with identical total numbers of individuals. One (call it A) occurs as a single large, panmictic population, the other (B) as a series of separate, fragmented populations. Which is most vulnerable to the proximate causes of extinction (Pimm 1991)? There is no simple answer, and indeed only vaguely useful contingent answers. It depends amongst other things, on the extent to which populations of B are isolated from one another, and the degree to which chance events strike populations of B independently or in synchrony (their spatial correlation).

If B's fragmented populations are totally isolated, then each is, on average, more vulnerable than the large population of A to the four causes of proximate extinction (Souleé 1987; Bengtsson 1993); however, it is unclear without knowing a great deal more about the system whether all the populations of B will die out before A. A counter argument, for example, is that a major catastrophe, such as disease or fire, may sweep through the range of A and wipe it out, whilst isolated populations of B may escape and survive for a time. More generally, catastrophies reduce the numerical advantages A has over populations of B (Mangel and Tier 1993). So there is no simple answer. Small, isolated, experimental *Daphnia* populations died out faster than equally isolated populations in larger pools in a replicated field experiment, but no disasters overtook the large experimental pools (Bengtsson 1993).

Total isolation is, however, extreme. Usually, metapopulations enjoy some exchange of individuals, allowing small populations of B to be 'rescued', and those that go extinct to re-establish via immigration. Is it now better to be A than B? Fragmentation into metapopulations increased extinction probabilities for laboratory stocks of *Drosophila pseudoobscura* and *D. hydei* (Forney and Gilpin 1989), and model populations reveal broadly similar results when extinctions are primarily due to demographic stochasticity (Burkey 1989). However, if environmental stochasticity is large and spatially uncorrelated, it may favour B, rather than A (Burkey 1993), because fragmentation spreads risk (den Boer 1981). (Note that no models currently incorporate the effects of fragmentation with all four causes of proximate extinction; if they did, genetic effects and social dysfunction could override the benefits of spreading of risk in metapopulations.)

A major problem in this debate is lack of knowledge about synchrony of fluctuations, driven by environmental stochasticity, in spatially separated conspecific populations. Increasing spatial synchrony decreases the expected lifetime of metapopulations (by reducing risk-spreading and thus making the separate populations of B behave, in aggregate, more like A) (e.g., Harrison and Quinn 1989). This problem has recently been reviewed and addressed by Hanski and Woiwod (1993), who show that for British moths and aphids, spatial synchrony between conspecific populations declines, as expected, with increasing distance between them, but synchrony remains positive at all distances up to 800 km. Spatial synchrony is itself positively correlated with temporal variability in local populations (except in geometrid moths); that is, the more similar fluctuations of conspecific populations are in space, the larger they are in time.

We may tentatively draw three conclusions. (1) Fragmented populations will have to be separated by many hundreds of kilometres before they can be guaranteed to fluctuate independently in response to environmental stochasticity. (2) Those populations most at risk from environmental stochasticity (i.e., those that fluctuate most) are apparently also at greatest risk of regional extinction because they show the greatest spatial synchrony; here is another double jeopardy. (3) Because spatial synchrony decreases the expected lifetime of metapopulations, risks of extinction are probably lower in a few large populations than in many small ones. Risks of inbreeding and social dysfunction are also lower in larger populations. Given a choice, current evidence therefore suggests that it is better to have all our population eggs in one or a few large geographical baskets (*A*) than many smaller ones (*B*).

Biology interacts with populations to change spatial patterns

It is unlikely that patterns of population decline and fragmentation are independent of the details of species' life histories and breeding systems. Is there any evidence, for instance, that some species have attributes that favour the persistence of very isolated populations, reducing the likelihood of extinction? Quinn *et al.* (in press) analysed the UK distributions of 139 species of scarce native British plants, mapped on a scale of 10 km × 10 km. The species examined occurred in 16–100 of these '10 km squares'. Quinn *et al.* refer to occupied 10 km squares as 'loci', because they are not strictly populations. Nevertheless, in the absence of any better information, the distributions of loci are instructive.

The loci of all 139 species are more aggregated than expected by chance, but some are more aggregated than others. Not surprisingly, habitat plays a significant role in determining patterns of aggregation at this scale, but so do two attributes of the plants themselves.

Species with seeds adapted for long-distance dispersal have less strongly aggregated loci than species with more feeble dispersal abilities, a result that is entirely consistent with metapopulation dynamic theory. In other words, fragmentation and isolation of populations poses a greater threat of extinction for species with poor dispersal abilities than for good dispersers, and it is the most isolated populations that go first.

More unexpectedly, obligate cross-pollinated plants have very few weakly aggregated species; they mainly have strongly aggregated loci. In contrast, species that can self-pollinate frequently have loci that are weakly aggregated. It is surprising that this result shows up at such a coarse scale, but it is consistent with the hypothesis that isolation of obligate outcrossers reduces their chance of reproductive success (Kunin 1992). Self-compatible plants may establish, and persist in, very isolated localities. (For further analyses of the characteristics of plants promoting persistence in small, isolated, and fragmented populations, see Bond, Chapter 9.)

A recent survey of the literature by Kunin and Gaston (1993) suggests that locally rare and geographically restricted species in many other taxa, from mosses and vascular plants to protozoa, insects, and mammals, may have characteristics that differ from more common relatives, including lower levels of self-incompatibility, and poorer dispersal abilities. As in Quinn *et al*.'s study, some of these characteristics may contribute to, and exacerbate one or more aspects of rarity, whilst others promote the persistence of rare populations. Whether any can be viewed as evolved adaptations to rarity (Kunin and Gaston 1993) and whether there are any differences between the characteristics of naturally rare versus anthropogenically rare species is a much more difficult question. I return to the problem briefly in the final section.

Summarizing, extant patterns of species' distributions—where populations persist after human impacts—are not immutable echoes of former ranges. Nor do all species react in the same way to similar abuses. Current ranges reflect biology and history, with some taxa being much more prone to local, regional, and ultimately global extinction than others. Although this seems very obvious, robust generalizations are scarce. A few, broad statistical patterns are now emerging, but there are many exceptions, and the generality of the patterns remains unclear.

10.4   Body size and trophic position

In a search for useful indicators of vulnerability to extinction, adult body size and trophic position might appear to offer some predictive power, at least for animals. Body size and trophic position are clearly correlated, but not perfectly, and common sense suggests that the ultimate causes of extinction driven by human activities differentially threaten larger herbivores and predators. Large animals need large home ranges (and so feel the effects of habitat destruction most keenly); they provide more meat per corpse, more excitement in the chase, and in some eyes more kudos when photographed dead or hung on a wall (and are therefore differentially hunted); they are perceived as direct competitors of, and a threat to people and livestock (and are persecuted accordingly); and predators, at the end of food chains, may accumulate more than their share of pesticides, heavy metals, and other pollutants. Being big is dangerous in a world dominated by *Homo sapiens*.

Despite these well-known concerns, for birds at least, body size (from 10 to 10 000 g) has no power to predict rates of population increase or decline in a global sample of threatened species from 12 families occupying various trophic levels (Green and Hirons 1991). I find these results surprising, and counterintuitive; it would be interesting to repeat the analyses controlling for generation time and taxonomy.

Body size as a predictor of proximate causes of extinction

Can we do better with the proximate causes of extinction? For example, if populations of two species differing in body weight by an order of magnitude both contain equal numbers of a few individuals, or if the two species are confined to a fixed area at natural densities, which population is most likely to persist, the larger- or the smaller-bodied one?

Again, there is no universally agreed answer, although the theoretical framework is reasonably well understood. Other things being equal, population persistence is more likely when average population size is large rather than small; when fluctuations in numbers are small, not large; and when recovery from low numbers is fast rather than slow (see also Pimm, Moulton, and Justice, Chapter 5). All these characteristics can be correlated with body size, but not necessarily in ways that act consistently to either promote or reduce the risk of extinction, and not, apparently in ways that are consistent across all taxa and body sizes. Space precludes a thorough summary of a complex literature, but some of the more pertinent points are as follows.

Although on a scale from invertebrates to whales, small-bodied species occur at higher average population densities than large-bodied species, densities vary by several orders of magnitude in organisms of one body size; and within local assemblages and communities, body size usually explains none, or virtually none, of the observed variation in average population numbers (Blackburn *et al.* 1993*a*; Blackburn and Lawton 1994). A knowledge of body size alone does not, therefore, allow us to predict which species will be vulnerable to extinction because they occur at low densities. We need to know numbers.

Second, although population variability seems a relatively easy concept, measuring it correctly is not simple (McArdle *et al.* 1990); links between observed variation in numbers and risk of population extinction are far from straightforward (McArdle and Gaston 1993); and population variability itself is correlated with body size in complex ways, both theoretically and empirically (Gaston and Lawton 1988; Pimm 1991; Sutherland and Baillie 1993). The best we can say is that on present evidence, and with some exceptions, populations of large-bodied species fluctuate less on an annual basis than smaller-bodied taxa (see Pimm 1991 for a review), making large-bodied species less likely to fluctuate unexpectedly to extinction in the short term.

One consequence of this broad generalization is that small, isolated populations of insects, endangered butterflies for example, are much more likely to fluctuate to extinction than equally isolated populations of birds of similar average population size, even when afforded maximum protection. Within more closely related taxa, populations of large-bodied species also appear less extinction-prone than related, small-bodied species, including *Daphnia* (Bengtsson 1993), shrews in the genus Sorex (Peltonen and Hanski 1991) and possibly some birds (Gotelli and Graves 1990).

However, more complex patterns have been claimed for local extinctions of birds on islands off Britain (Pimm *et al.* 1988). This analysis suggest that below a population size of about 7 breeding pairs, the probability of extinction is smaller for large-bodied species than small-bodied species, with the risks reversing in larger populations. A re-analysis does not confirm these results (Tracy and George 1992); I have sympathy and disagreements with both sides of this debate, and believe that the matter is still unresolved.

Trophic position

Vulnerability to local extinction is unlikely to be independent of where species fit into food chains. For example, simple models predict that populations at the top of long food chains are more likely to become extinct than those at lower trophic levels in the noisy, real world (Pimm and Lawton 1977); and that intermediate species that suffer both predation and competition from higher-level 'omnivorous' predators ('intra-guild predation') are vulnerable to extinction (Pimm and Lawton 1978; Polis *et al.* 1989). Both predictions have some support (e.g., Schoener and Spiller 1987; Spiller and Schoener 1990; Jenkins *et al.* 1992). Pimm (1991) reviews the whole problem of how trophic position may influence population dynamics and persistence.

The problem is to distinguish trophic position *per se* from other confounding variables, not least body size (which tends to be bigger at higher trophic levels in predator food chains), and other natural history details (some of which will also be correlated with body size). The prediction that higher trophic levels are more vulnerable to environmental stochasticity, for example (see previous paragraph), appears inconsistent with the claim for greater population persistence in larger-bodied species (made in section 10.3), unless other things are factored out. Data and experiments that address these problems are very few.

One, pioneer experiment bears examination. In small aquatic microcosms with protozoan predators and bacterial prey, involving various combinations of species in food webs of different complexity, prey went extinct more often than predator populations and extinctions were associated with increasing species-richness (Lawler 1993). But these results were sensitive to exactly which species were in the microcosms; as Lawler notes, the characteristics of individual species matter as much as the abstract trophic web in which they sit.

Perhaps for these reasons, and contrary to expectations, it has proved extremely difficult to detect a consistent tendency for differential extinction of species at higher trophic levels, either in the fossil record (Jablonski 1986 and Chapter 2) or as a consequence of habitat fragmentation in extant communities (Mikkelson 1993 and references therein). Mikkelson's analyses suggest that all trophic levels lose species pro rata, so that proportions of species in different trophic categories remain constant as diversity falls.

10.5 Phylogenetic constraints

Two of the most important variables that determine risks of extinction, namely, population density and size of geographic range, are usually thought of as labile attributes of species, set by processes operating in ecological time. That is, in the absence of major impacts by humans, we expect abundances particularly, and ranges occasionally, to be dynamic and variable over time periods of, say 10 to 10^3 years. But there are also poorly understood, intriguing hints of effects operating in evolutionary time. Obviously, individuals, not species, are the units of selection; nevertheless, data now exist suggesting that both range and abundance are persistent, species' characteristics. The idea is implicit in Brown's hypothesis (1984), which links local abundance and size of range to a complex, elusive, albeit obviously evolved characteristic of species, namely fundamental niche breadth. One implication of these studies is that some taxa are more extinction-prone than others.

The data are summarized by Lawton (1993). One group of studies centres on the 'taxon cycle' for birds on West Indian islands (Ricklefs and Cox 1978; Ricklefs 1989). Among passerines, putatively older taxa occur on fewer islands, have more restricted habitat distributions, and tend to have reduced population densities. They would appear to be particularly sensitive to human impact.

A second group of studies also involves birds, and finds quite unexpected phylogenetic effects on patterns in species' abundances as a function of body size. Briefly summarizing a complex literature, body size:abundance plots for species within individual tribes of birds show significantly more *positive* relationships than expected (large-bodied species are more common than small-bodied species) in taxonomically more ancient tribes (Cotgreave and Harvey 1991; Nee *et al.* 1991; Blackburn *et al.* 1994). This pattern is not well understood, but at least in part it could be the product of differential extinctions of large-bodied, rare species over long periods of time. (Although this suggestion sits uncomfortably with some of the data and theory discussed earlier on pp. 156–7.)

Plants also provide evidence for phylogenetic effects on distribution and abundance. Rare North American plants (loosely defined by range, local abundance, or both) are significantly over-represented in certain families (Scrophulariaceae, Lamiaceae) and under-represented in others (Rosaceae) (Schwartz 1993). More unexpectedly, disjunct taxa within extant genera of herbaceous perennial plants, relict in temperate eastern Asia and eastern North America, have significantly correlated range sizes, implying stasis in genus-level attributes determining distributions that have been stable for at least 10 million years (Ricklefs and Latham 1992). Unfortunately, we have no idea what characteristics of these taxa promote commonness or rarity.

Phylogenetic constraints on range sizes have also been reported for animals. Individual species of bivalves and gastropods from late Cretaceous fossil assemblages in North America achieved characteristic range sizes relatively early in their history; once evolved, species' range sizes changed relatively little. Moreover, pairs of closely related species have statistically similar range sizes (Jablonski 1987). Again, why this should be is unclear, but the fossil record also shows that locally endemic species with small geographic ranges have a much higher chance of extinction than more cosmopolitan genera (Raup and Jablonski 1993; Jablonski, Chapter 2).

If we accept that these patterns are real (which is not certain) they mean two things. First, some taxa (genera, tribes, families) are more extinction-prone than others, but we have no idea why. Second, species phylogenetically predisposed to low population densities and small geographic ranges will be unusually vulnerable to the ultimate causes of extinction, in the form of exploitation and habitat destruction. Paradoxically, however, they may also be *less* prone to the proximate causes of extinction because they are pre-adapted to life as small populations on isolated reserves, compared with formerly more abundant and widespread species (see also Kunin and Gaston 1993, p. 155). The problem deserves more urgent attention.

Summary

Although the causes of population extinction, locally and globally, are many, varied and idiosyncratic when individual species are considered, there are nevertheless some basic general constraints and ground rules that make some species more extinction-prone than others. It is useful, first of all, to distinguish between ultimate causes of extinction—why species are, or become, rare in the first place; and proximate causes—why small populations may still die out, even when protected.

Species most at risk from proximate causes of extinction are those with small geographic ranges and small total populations; ironically range and population density tend to be positively correlated, putting some populations in 'double jeopardy'. Tropical taxa may also have smaller average range sizes than related, temperate taxa, increasing their vulnerability to habitat destruction. Links between geographic range, metapopulation processes, and patterns of population decline deserve more attention.

Other correlates of extinction risk include body size and trophic position, but here generalizations are less easy to make. Large-bodied species are particularly vulnerable to ultimate causes of extinction (because of hunting, or habitat destruction, for instance), but may be less at risk to proximate causes (because their populations fluctuate less). On present evidence, and unexpectedly, trophic position has no consistent effect on the risk of extinction from proximate causes.

Finally, there is growing evidence that some higher taxa (families, orders, etc.) are more extinction-prone than others. Although the reasons are often poorly understood, the implications are stark; direct and indirect human destruction of species is about to become a major force in shaping the future of life on Earth. Those species and higher taxa that survive to the end of the next century and beyond will not be a random sample of extant organisms. We may not wish to destroy gorillas and bowerbirds and encourage starlings and cockroaches. But we seem powerless to do anything about it.

Acknowledgements

Bob May made helpful comments on the manuscript. Work during the preparation of this chapter was supported by the core grant to the NERC Centre for Population Biology.

References

Andrewartha, H. G. and Birch, L. C. (1954). *The distribution and abundance of animals.* University of Chicago Press.

Bengtsson, J. (1993). Interspecific competition and determinants of extinction in experimental populations of three rockpool *Daphnia* species. *Oikos,* **67**, 451–64.

Blackburn, T. M. and Lawton, J. H. (1994). Population abundance and body size in animal assemblages. *Phil. Trans. Roy. Soc. Lond.,* **B343**, 33–9.

Blackburn, T. M., Brown, V. K., Doube, B. M., Greenwood, J. J. D., Lawton, J. H., and Stork, N. E. (1993*a*). The relationship between abundance and body size in natural animal assemblages. *J. Anim. Ecol.,* **62**, 519–28.

Blackburn, T. M., Gates, S., Lawton, J. H., and Greenwood, J. J. D. (1994). Relationships between body size, abundance, and taxonomy of birds wintering in Britain and Ireland. *Phil. Trans. Roy. Soc. Lond.,* **B343**, 135–44.

Brown, J. H. (1984). On the relationship between abundance and distribution of species. *Amer. Nat.,* **124**, 255–79.

Burkey, T. V. (1989). Extinction in nature reserves: the effect of fragmentation and the importance of migration between reserve fragments. *Oikos,* **55**, 75–81.

Burkey, T. V. (1993). Living dangerously but independently, or safely and contingently? *TREE,* **8**, 302.

Caughley, G. (1994). Directions in conservation biology. *J. Anim. Ecol.,* **63**, 215–44.

Caughley, G., Grice, D., Barker, R., and Brown, B. (1988). The edge of range. *J. Anim. Ecol.,* **57**, 771–85.

Cotgreave, P. and Harvey, P. H. (1991). Bird community structure. *Nature,* **353**, 123.

Cracraft, J. (1992). Explaining patterns of biological diversity: integrating causation at different spatial and temporal scales. In *Systematics, ecology, and the biodiversity crisis,* (ed. N. Eldredge), pp. 59–76. Columbia University Press, New York.

Cramp, S. and Simmons, K. L. (ed.) (1980). *Handbook of the birds of Europe, the Middle East, and North Africa,* Vol. 2. *Hawks to bustards.* Oxford University Press.

Currie, D. J. and Fritz, J. T. (1993). Global patterns of animal abundance and species energy use. *Oikos,* **67**, 56–68.

den Boer, P. J. (1981). On the survival of populations in a heterogeneous and variable environment. *Oecologia,* **50**, 39–53.

Forney, K. A. and Gilpin, M. E. (1989). Spatial structure and population extinction: a study with *Drosophila* flies. *Conserv. Biol.,* **3**, 45–51.

France, R. (1992). The North American latitudinal gradient in species richness and geographic range of freshwater crayfish and amphipods. *Amer. Nat.,* **139**, 342–54.

Gaston, K. J. (1991). How large is a species' geographic range? *Oikos,* **61**, 434–8.

Gaston, K. J. and Lawton, J. H. (1988). Patterns in the distribution and abundance of insect populations. *Nature,* **331**, 709–12.

Gaston, K. J. and Lawton, J. H. (1990). Effects of scale and habitat on the relationship between regional distribution and local abundance. *Oikos,* **58**, 329–35.

Gotelli, N. J. and Graves, G. R. (1990). Body size and the occurrence of avian species on land-bridge islands. *J. Biogeog.,* **17**, 315–25.

Green, R. E. and Hirons, G. J. M. (1991). The relevance of population studies to the conservation of threatened birds. In: *Bird population studies. Relevance to conservation and management,* (ed. C. M. Perrins, J.-D. Lebreton, and G. J. M. Hirons), pp. 594–633. Oxford University Press.

Griffith, B., Scott, J. M., Carpenter, J. W., and Reed, C. (1989). Translocation as a species conservation tool: status and strategy. *Science,* **245**, 477–80.

Gyllenberg, M. and Hanski, I. (1992). Single-species metapopulation dynamics: a structured model. *Theoret. Pop. Biol.,* **42**, 35–66.

Hanski, I. (1982). Dynamics of regional distribution: the core and satellite hyothesis. *Oikos*, **38**, 210–21.

Hanski, I. and Woiwod, I. P. (1993). Spatial synchrony in the dynamics of moth and aphid populations. *J. Anim. Ecol.*, **62**, 656–68.

Hanski, I., Kouki, J., and Halkka, A. (1994). Three explanations of the positive relationship between distribution and abundance of species. In *Species diversity in ecological communities: Historical and geographical perspectives*, (ed. E. Ricklefs and D. Schulter), pp. 108–16. University of Chicago Press.

Harrison, S. (1991). Local extinction in a metapopulation context: an empirical evaluation. *Biol. J. Linn. Soc.*, **42**, 73–88.

Harrison, S. and Quinn, J. F. (1989). Correlated environments and the persistence of metapopulations. *Oikos*, **56**, 293–8.

Harrison, S., Murphy, D. D., and Ehrlich, P. R. (1988). Distribution of the bay checkerspot butterfly, *Euphydryas editha bayensis*: evidence for a metapopulation model. *Amer. Nat.*, **132**, 360–82.

Hengeveld, R. (1989). *Dynamics of biological invasions*. Chapman & Hall, London.

Hengeveld, R. and Haeck, J. (1982). The distribution of abundance. I. Measurements. *J. Biogeog.*, **9**, 303–6.

Huffaker, C. B. and Messanger, P. S. (1964). The concept and significance of natural control. In *Biological control of insect pests and weeds*, (ed. P. De Bach), pp. 74–117. Chapman & Hall, London.

Jablonski, D. (1986). Larval ecology and macroevolution in marine invertebrates. *Bull. Mar. Sci.*, **39**, 565–87.

Jablonski, D. (1987). Heritability at the species level: analysis of geographic ranges of Cretaceous mollusks. *Science*, **238**, 360–3.

Jenkins, B., Kitching, R. L., and Pimm, S. L. (1992). Productivity, disturbance and food web structure at a local spatial scale in experimental container habitats. *Oikos*, **65**, 249–55.

Kunin, W. (1992). Density and reproductive success in wild populations of *Diplotaxis erucoides* (Brassicaceae). *Oecologia*, **91**, 129–33.

Kunin, W. E. and Gaston, K. J. (1993). The biology of rarity: patterns, causes and consequences. *TREE*, **8**, 289–301.

Lande, R. (1991). Population dynamics and extinction in heterogeneous environments: the Northern Spotted Owl. In *Bird population studies. Relevance to conservation and management*, (ed. C. M. Perrins, J.-D. Lebreton, and G. J. M. Hirons), pp. 566–80. Oxford University Press.

Lawler, S. P. (1993). Species richness, species composition and population dynamics of protists in experimental microcosms. *J. Anim. Ecol.*, **62**, 711–19.

Lawton, J. H. (1993). Range, population abundance and conservation. *TREE*, **8**, 409–13.

Lawton, J. H., Nee, S., Letcher, A. J., and Harvey, P. H. (1994). Animal distributions: patterns and processes. In *Large-scale ecology and conservation biology*, (ed. P. J. Edwards, R. M. May, and N. R. Webb), pp. 41–58. Blackwell, Oxford.

Mangel, M. and Tier, C. (1993). A simple direct method for finding persistence times of populations and application to conservation problems. *PNAS*, **90**, 1083–6.

McArdle, B. and Gaston, K. (1993). The temporal variability of populations. *Oikos*, **67**, 187–91.

McArdle, B. H., Gaston, K. J., and Lawton, J. H. (1990). Variation in the size of animal populations: patterns, problems and artefacts. *J. Anim. Ecol.*, **59**, 439–54.

Mikkelson, G. M. (1993). How do food webs fall apart? A study of changes in trophic structure during relaxation on habitat fragments. *Oikos*, **67**, 539–47.

Nee, S., Read, A. F., Greenwood, J. J. D., and Harvey, P. H. (1991). The relationship between abundance and body size in British birds. *Nature*, **351**, 312–13.

Peltonen, A. and Hanski, I. (1991). Patterns of island occupancy explained by colonization and extinction rates in shrews. *Ecology*, **72**, 1698–1708.

Pimm, S. L. (1991). *The balance of nature?* University of Chicago Press.

Pimm, S. L. and Lawton, J. H. (1977). Number of trophic levels in ecological communities. *Nature*, **268**, 329–31.

Pimm, S. L. and Lawton, J. H. (1978). On feeding on more than one trophic level. *Nature*, **275**, 542–4.

Pimm, S. L., Jones, H. L. and Diamond, J. (1988). On the risk of extinction. *Amer. Nat.*, **132**, 757–85.

Polis, G. A., Myers, C. A., and Holt, R. D. (1989). The ecology and evolution of intraguild predation: potential competitors that eat each other. *Ann. Rev. Ecol. Syst.*, **20**, 297–330.

Pulliam, H. R. (1988). Sources, sinks, and population regulation. *Amer. Nat.*, **132**, 652–61.

Quinn, R. M., Lawton, J. H., Eversham, B. C., and Wood, S. N. The biogeography of scarce vascular plants in Britain with respect to habitat preference, dispersal ability and reproductive biology. *Biol. Conserv.*, in press.

Rabinowitz, D., Cairns, S., and Dillon, T. (1986). Seven forms of rarity and their frequency in the flora of the British Isles. In *Conservation biology. The science of scarcity and diversity*, (ed. M. E. Souleé), pp. 182–204. Sinauer, Sunderland, MA.

Randall, M. G. M. (1982). The dynamics of an insect population throughout its altitudinal distribution: *Coleophora alticolella* (Lepidoptera) in northern England. *J. Anim. Ecol.*, **51**, 993–1016.

Rands, M. R. W. (1991). Conserving threatened birds: an overview of the species and the threats. In *Bird population studies: Relevance to conservation and management*, (ed. C. M. Perrins, J.-D. Lebreton, and G. J. M. Hirons), pp. 580–93. Oxford University Press.

Rapoport, E. H. (1982). *Areography: Geographical strategies of species*. Pergamon, Oxford.

Raup, D. M. and Jablonski, D. (1993). Geography of end-Cretaceous marine bivalve extinctions. *Science*, **260**, 970–3.

Richards, O. W. (1961). The theoretical and practical study of natural insect populations. *Ann. Rev. Entomol.*, **6**, 147–62.

Ricklefs, R. E. (1989). Speciation and diversity: the integration of local and regional processes. In: *Specification and its consequences*, (ed. D. Otte and J. A. Endler), pp. 599–622. Sinauer, Sunderland, MA.

Ricklefs, R. E. and Cox, G. W. (1978). Stage of taxon cycle, habitat distribution, and population density in the avifauna of the West Indies. *Amer. Nat.*, **112**, 875–95.

Ricklefs, R. E. and Latham, R. E. (1992). Intercontinental correlation of geographical ranges suggests stasis in ecological traits of relict genera of temperate perennial herbs. *Amer. Nat.*, **139**, 1305–21.

Rogers, D. J. and Randolph, S. E. (1986). Distribution and abundance of tsetse flies (*Glossina* spp.). *J. Anim. Ecol.*, **55**, 1007–25.

Root, T. (1988). *Atlas of wintering North American birds*. University of Chicago Press.

Schoener, T. W. and Spiller, D. A. (1987). Effects of lizards on spider populations: manipulative reconstruction of a natural experiment. *Science*, **236**, 949–52.

Schwartz, M. W. (1993). The search for patterns among rare plants: are primitive species more likely to be rare? *Biol. Conserv.*, **64**, 121–7.

Simberloff, D. (1986). The proximate causes of extinction. In *Patterns and processes in the history of life*, (ed. D. M. Raup and D. Jablonski), pp. 259–76. Springer, Berlin.

Souleé, M. E. (ed.) (1987). *Viable populations for conservation*. Cambridge University Press.

Spiller, D. A. and Schoener, T. W. (1990). Lizards reduce food consumption by spiders: mechanisms and consequences. *Oecologia*, **83**, 150–61.

Stevens, G. C. (1989). The latitudinal gradient in geographic range: how so many species coexist in the tropics. *Amer. Nat.*, **133**, 240–56.

Sutherland, W. J. and Baillie, S. R. (1993). Patterns in the distribution, abundance and variation of bird populations. *Ibis*, **135**, 209–10.

Taylor, R. A. J. and Taylor, L. R. (1979). A behavioural model for the evolution of spatial dynamics. In *Population dynamics. 20th Symposium of the British Ecological Society*, (ed. R. M. Anderson, B. D. Turner, and L. R. Taylor), pp. 1–27. Blackwell, Oxford.

Thomas, C. D. and Jones, T. M. (1993). Partial recovery of a skipper butterfly (*Hesperia comma* from population refuges: lessons for conservation in a fragmented landscape. *J. Anim. Ecol.*, **62**, 472–81.

Tracy, C. R. and George, T. L. (1992). On the determinants of extinction. *Amer. Nat.*, **139**, 102–22.

van Wilgen, B. W., Bond, W. J., and Richardson, D. M. (1992). Ecosystem management. In *The ecology of fynbos. Nutrients, fire, and diversity*, (ed. R. M. Cowling), pp. 345–71. Oxford University Press.

Wiens, J. A. (1989). *The ecology of bird communities. Foundations and patterns*, Vol. 1. Cambridge University Press.

Wilcove, D. S. and Terborg, J. W. (1984). Patterns of population decline in birds. *Amer. Birds*, **38**, 10–13.

11

Estimating extinction from molecular phylogenies

Sean Nee, Edward C. Holmes, Robert M. May, and Paul H. Harvey

11.1 Introduction

Increasing mechanization is turning a trickle of molecular information about phylogenies into a flood. Molecular phylogenies can both describe the hierarchical relationships among taxa by defining clades (monophyletic groups), and provide a time axis by utilizing molecular clocks. Even when those clocks have not been calibrated against real time, molecular phylogenies can still describe the temporal orderings of nodes relative to each other. We have shown elsewhere how many properties of the past, present, and even the future of a clade leave their footprints in its molecular phylogeny, even if that phylogeny is based only on information from extant organisms (Harvey *et al.* 1991, in press *b*; Nee *et al.* 1992). As a consequence, it is possible to use a careful combination of theoretical modelling, statistical analysis, and biological knowledge of the clade under investigation to make inferences about many aspects of the tempo and mode of its evolutionary history. In this chapter we describe how such inferences can be made, and produce illustrative examples to show how extinction rates can be estimated and other components of biodiversity can be measured.

The chapter is organized in the following way. We first describe the relationship between the *actual* phylogeny of a group, as would have been recorded in a perfect fossil record, and its molecular phylogeny, which we will refer to as the *reconstructed* phylogeny as it is based solely on extant species. Utilizing this distinction, we show how the reconstructed phylogeny can be used to determine which clades may have enjoyed unusually high rates of cladogenesis (lineage splitting) and to identify biological correlates of diversification. We then introduce a simple model of cladogenesis, which has a good pedigree in palaeontology (Gould *et al.* 1977; Raup *et al.* 1973; Stanley 1979), the constant rate birth–death process, and explore its implications for the analysis of the reconstructed phylogeny of a complete clade. We extend that model to describe situations in which the molecular phylogeny is based on only a random sample of extant species from the clade, and show with illustrative examples how that extended model can provide important information for estimating biodiversity. Finally, to demonstrate how the interpretation of molecular phylogenies requires the consideration of a variety of evolutionary scenarios, we describe how a model which assumes that the number of lineages remains constant through time produces a phylogenetic tree with a radically different structure from that produced by a model in which the number of lineages has been increasing through time, as in the constant rate birth–death model. -

Important conclusions to be drawn from this chapter will be that the appearance of reconstructed phylogenies can be both counterintuitive and deceptive with regard to the actual processes that have generated them, and that evolutionarily different processes can leave distinctively different marks on the reconstructed phylogeny.

11.2   Actual and reconstructed phylogenies

We first distinguish between actual and reconstructed phylogenies. Consider Fig. 11.1 which shows an actual phylogeny with lineages that gave rise to descendants in the present day picked out in bold. The bold-lined phylogeny, with kinks removed, is the reconstructed phylogeny. We note four points when examining the actual and reconstructed phylogenies. First, both phylogenies have the same number of taxa in the present day. Second, at any point in the past, the reconstructed phylogeny generally has fewer lineages present (and never more) than does the actual phylogeny. Third, while the number of lineages can decrease towards the present in the actual phylogeny, that cannot happen in the reconstructed phylogeny. Finally, the reconstructed phylogeny provides timings for when each pair of species last shared a common ancestor and, therefore, commences at that point in the past when all present-day species shared their most recent common ancestor.

The first observation that strikes people with an interest in biodiversity is the enormous extent to which diversity varies amongst taxa—the huge difference between the number of species of beetles and butterflies would be a good example. In making a decision whether or not the causes of a clade's apparently high diversification should be investigated, it is desirable to know whether or not the diversification really *is* remarkable with reference to some null model.

Fortunately, such an analysis is very simple. Under a birth–death model (see below), the number of progeny of any particular lineage in a reconstructed phylogeny has a geometric distribution, that is, the probability of having n progeny some time later is given by the expression $(1-u)u^{n-1}$, where the single parameter of the distribution, u, is determined by the amount of time later and by the birth and death parameters. This is a very general result (Nee *et al.* 1994). It does not require that birth and death rates are constant—they may vary over time. In fact, it does not even depend on the phylogeny actually being generated by a birth–death model. Once we have made the central assumption of our null hypothesis that, as we look through time in our molecular phylogeny, the next bifurcation is as likely to occur on any lineage as on any other, then the progeny number distribution of the lineages is geometric. (This is because, by making this assumption, we have just defined a pure birth process, which can be used to

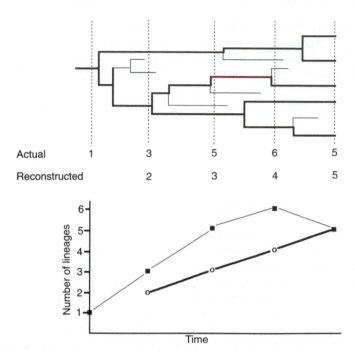

Actual	1	3	5	6	5
Reconstructed		2	3	4	5

Fig. 11.1   A hypothetical phylogeny as it would appear in a perfect fossil record. The bold lines represent those lineages which have some descendants at the present day and, so, would appear in a phylogeny reconstructed from molecular data. The numbers of lineages through time for the actual and reconstructed phylogenies are plotted at the bottom of the figure.

generate any phylogeny, if the birth rate is defined appropriately as a function of time.) Finally, we do not need all the lineages—a random sample will do.

So, with the geometric distribution in hand, how do we go about identifying clades that are unusually bushy, where 'unusually' will be defined in a precise way as 'statistically anomalous'? Consider Fig. 11.2 where we have drawn two vertical lines across a hypothetical reconstructed phylogeny (the right-hand line could correspond to the present). We now ask the question, for each lineage crossing the left-hand line, how many progeny does it have, including itself, by the time defined by the right-hand line? If there are k lineages at any particular time and n lineages sometime later, then all vectors of progeny numbers $(x_1, x_2, \ldots x_k)$, such that the sum total of the elements is n, are equally probable. This follows immediately from the geometric distribution of progeny number, since the probability of any vector of progeny numbers is proportional to $(1-u)^k u^{n-k}$. So the statistics of progeny number distribution can be determined by randomly breaking a stick n units long into k fragments, and allowing breakages to occur only at unit boundaries. The lengths of the fragments correspond to progeny numbers and each broken stick corresponds to an equiprobable vector of

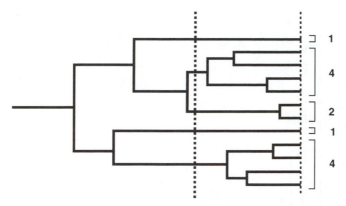

Fig. 11.2   A hypothetical reconstructed phylogeny with two lines drawn across it. For each lineage crossing the left-hand line, we can ask how many progeny lineages, including itself, it has by the time defined by the right-hand line.

progeny numbers. It is straightforward to repeatedly break sticks on a computer and, thereby, determine whatever statistics are of interest.

As an example of this procedure, Fig. 11.3 shows the results of an analysis of the molecular phylogeny of the birds constructed by Sibley and Ahlquist (Harvey *et al.* 1991; Nee *et al.* 1992). The histogram shows the numbers of lineages with 1, 2, 3, etc. progeny lineages at the later time chosen for the analysis, and the curve shows the fitted geometric distribution. The two lineages that gave rise to 15 and 19 progeny lineages, the Passeri and Ciconiiformes, respectively, certainly appear to be anomalous and statistical analysis confirms this: the probability in this case of any lineage giving rise to more than 14 progeny lineages is less than 0.005.

In addition to identifying clades which are unusually diverse, we can also exploit molecular phylogenies to construct parametric tests of hypotheses concerning biological correlates of diversification. One way to do this is as follows. To focus the discussion, we will suppose that we are testing the hypothesis that small body size promotes cladogenesis, an hypothesis that has been put forward to account for the radiation of the passerines. Having reconstructed ancestral character states, we can ask, for each node in a phylogeny, whether the branch leading to the smaller-bodied clade is longer or shorter than the branch leading to the larger-bodied clade (Nee *et al.* 1992). If we assume that the lengths of the branches from a node are drawn from an exponential distribution, then the variance of the logarithmically transformed branch lengths is $\psi^{(1)}(1)$ ($=\pi^2/6$), where $\psi^{(1)}$ is the trigamma function (Cox and Lewis 1966). (The reasonableness of the exponential distribution assumption is determined by the data.) Under the null hypothesis that the branch lengths leading to the lighter and heavier daughter clades have the same expectation, the difference in the logarithmically transformed lengths has an expectation of zero and a variance of $2\psi^{(1)}(1)$. The sum of N such differences, where N is large, is approximately normally distributed with an expectation of zero and variance of

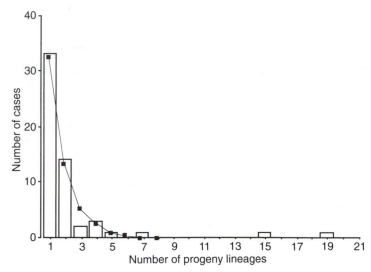

Fig. 11.3 The results of the time window analysis, described in the text and Fig. 11.2, as applied to the Sibley and Ahlquist's molecular phylogeny of the birds. The histogram shows the numbers of lineages with 1, 2, 3, etc. progeny lineages, and the line is the expectation from the fitted geometric distribution.

$2N\psi^{(1)}(1)$. Transformation now allows a Z-test to be performed. Notice that the null hypothesis does not assume that the expected branch lengths are the same all over the tree, merely the same for the two branches derived from each node. We performed this analysis on Sibley and Ahlquist's (1990) phylogeny of the birds, and found no evidence to support the hypothesis that body size promotes cladogenesis (Nee *et al.* 1992).

Another approach to testing for correlates of diversification uses sister taxon comparisons, comparing the sizes of the two clades on each side of a node. This sort of analysis (Slowinski and Guyer 1993) exploits, once again, the geometric distribution arising from the null hypothesis.

11.3 The constant rate birth–death model

In order to make further inferences about various quantities of interest from the evidence provided by reconstructed phylogenies, it is necessary to have a variety of explicit theoretical models which describe how these phylogenies may have been generated. We will now consider one of the simplest, the constant rate birth–death process. This model supposes that each lineage, at each point in time, has the same probability of giving rise to a new lineage (birth), or of going extinct (death), as any other lineage, and that these probabilities do not change over time. The growth of reconstructed phylogenies under this model, and its generalizations, is analysed elsewhere (Nee *et al.* 1994). In order to describe its properties in broad terms, we will represent the information contained in a

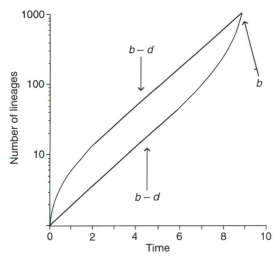

Fig. 11.4 The theoretically expected growth in the numbers of lineages through time for an actual (top line) and reconstructed phylogeny growing according to a constant rate birth–death process. The slopes of both curves are $b-d$, the speciation rate minus the extinction rate, over most of the history of the clade, and the slope of the reconstructed phylogeny asymptotically approaches the speciation rate towards the present day. The two curves pull apart further the greater the ratio of the extinction rate to the speciation rate.

reconstructed phylogeny simply as a lineage-through-time plot (Fig. 11.4), in which points are plotted when the second, third, fourth, etc. lineages appear and we then connect the dots (Harvey *et al.* in press *b*).

If there was no extinction (i.e., the phylogeny grew as a pure birth process), then the curve representing the actual number of lineages through time in Fig. 11.4 would be exactly coincident with the curve representing the number through time in the reconstructed phylogeny, and this would be a straight line on the semilogarithmic plot, since, on such a plot, the slope of the line corresponds to the per lineage rate of cladogenesis. As the death rate d becomes larger relative to the birth rate b (but still $d<b$), the two lines pull apart, remaining joined solely at the origin and the present day. The slopes of the lines are the same most of the time, equal to the birth rate minus the death rate $(b-d)$. Each line has a period of curvature—which we call 'the push of the past' and the 'pull of the present'. The push of the past, the apparently higher rate of cladogenesis at the beginning of the growth of the actual phylogeny, results the fact that we are only considering those clades which survived to the present day (so we can perform a molecular phylogenetic analysis), and these are the ones which on average got off to a 'flying start'. The pull of the present, the apparent increase in the rate of cladogenesis in the recent past in the reconstructed phylogeny, results from the fact that lineages which arose in the more recent past have had less time to go extinct and, so, are more likely to be represented in our reconstructed phylogeny. In fact, the slope of

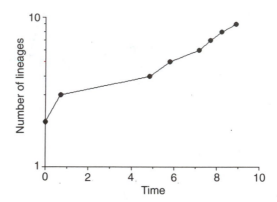

Fig. 11.5   The lineages-through-time plot for species of the *Drosophila melanogaster* subgroup.

the reconstructed curve asymptotically approaches the birth rate as we get closer to the present (Harvey *et al.* in press *b*). These effects, the pull of the present and the push of the past, are controlled by the ratio of the death rate to the birth rate, d/b, becoming larger as d/b increases towards unity.

In fact, the parameters of most interest for phylogenies which grow according to this model are not b and d separately, but functions of b and d; specifically, $(b-d)$ and d/b. The first, $(b-d)$, controls the rate of growth of the phylogeny and the second, d/b, controls both the magnitude of the pull of the present and the vulnerability of small clades to extinction through 'demographic stochasticity' (*sensu* May 1973; see MacArthur and Wilson 1967).

Consider, now, Fig. 11.5, which shows the increase through time in the number of lineages in a molecular phylogeny of the *Drosophila melanogaster* species subgroup constructed by Caccone *et al.* (1988). The graph starts when there are two lineages, because we have no information about the time of origin of the first lineage, and the time axis is in arbitrary units. The curve appears to be a straight line, with stochastic wobbling, and simple inspection suggests that a constant rate birth–death model is a reasonable assumption for the data, and that the extinction rate for this group is, in fact, zero, since there is no upward curve towards the present.

We can do better than this, however. Using the underlying probability model, we can construct a log likelihood surface for the parameters (Fig. 11.6), from which we may make inferences about what parameter values are supported by the data. The height of the surface at any point is the logarithm of the probability of the data, given the parameter values at that point. The peak of this surface, marked by an X, corresponds to the maximum likelihood estimate. Consistent with our visual impression, the maximum likelihood estimate of d/b is zero (i.e., a zero rate of extinction). Following the procedures that are conventionally used in such plots, the contour lines shown in the figure correspond to one, two, and three units of 'support', where support is the difference in height between the

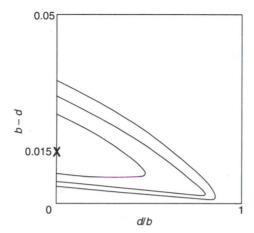

Fig. 11.6   Contour plot of the likelihood surface for the *Drosophila* data. The maximum
likelihood estimate, the peak of the surface, is marked by an X.

maximum likelihood point of the surface and elsewhere. The maximum
likelihood estimate of the parameters is about 7 times more likely to produce the
observed data than any point on the second contour line, and 20 times more
likely to produce the data than any point on the third contour line.

The interpretation of likelihood surfaces sees philosophy rear its ugly head. If
one subscribes to an extreme 'likelihood' approach, such as advocated by
Edwards (1972), then 'the interpretation of the support surface lies with the
individual investigator' (Thompson 1975). For this school of thought, by
convention, the second contour line in the figure corresponds, in some very loose
sense, to the notion of a confidence interval for the parameters. If, on the other
hand, one subscribes to the 'strong repeated sampling principle' (see Edwards
(1972) or Cox and Hinkley (1992) for a discussion of these issues), then one can, in
principle, put a conventional confidence interval around the maximum
likelihood estimate of the parameters. The '$P < 0.05$' confidence contour would
then lie somewhere between the second and third contour line shown in Fig. 11.6,
if the χ^2 approximation to the likelihood ratio statistic is reasonably good. (We
do not show this contour, because it would suggest a degree of precision about its
location which is unjustified.)

Philosophy aside, the shallowness of the likelihood surface in the d/b direction
tells us quite clearly that we cannot exclude the possibility that, in fact, *Drosophila*
has a substantial value of d/b. Although *Drosophila* give the *appearance* of having
a zero extinction rate, this analysis shows us that we cannot have any great degree
of confidence in that conclusion on the basis of these data alone. Such uncertainty
about the parameters for small clades is not entirely a weakness of the evidence
provided by reconstructed phylogenies as opposed to actual phylogenies: Fig.
11.7 shows the likelihood surface for a simulated actual phylogeny with about the
same number of species at the present day as the *Drosophila* subgroup. One

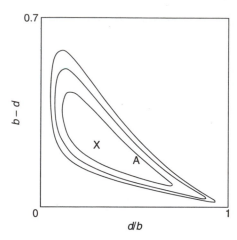

Fig. 11.7   Contour plot of the likelihood surface for a simulated actual phylogeny of a clade that grows to the same size as the *Drosophila* clade. The parameters chosen for the simulation were as follows: $b=0.4$, $d=0.2$, and the simulation was run for 20 time units. The letter A denotes the actual parameter values used and the X denotes the maximum likelihood estimate.

benefit of this likelihood-surface approach, as opposed to simply identifying an estimate of a quantity of interest, is that it quantifies our uncertainty: knowledge of what is uncertain is knowledge gained.

Consider now another illustrative example shown in Fig. 11.8: the increase through time in the number of lineages of salamanders of the genus *Plethodon*. The data are taken from a molecular phylogeny constructed by Highton and Larson (1979), who rightly drew attention to the apparent rapid acceleration in the rate of cladogenesis in the recent past in this group. In the absence of theoretical analysis, it would be tempting to conclude that it is 'salad days for salamanders' (Angela MacLean, pers. comm.) and that the future of the genus looks rosy. But appearances can deceive. In fact, we know that an apparent rapid acceleration in the rate of cladogenesis in the recent past is a property of reconstructed phylogenies that grow according to a birth–death process with a death rate that is close to the birth rate. So, in fact, it may be more reasonable to conclude that *Plethodon* has a large d/b, and, so, is highly vulnerable to extinction. This impression is confirmed by the likelihood-surface analysis (Fig. 11.9) which, again, leaves us with much uncertainty about the parameter estimates, but shows us that the best estimate for d/b is quite large.

Of course, one might choose to compare the performance of a constant rate model with a variable rate model on the *Plethodon* data in order to decide whether or not the increase in the rate of cladogenesis is real or apparent. But we are not, here, analysing whether or not a model is a good description of the data (we shall perform such an analysis below), or comparing the performances of different models (which would be done using standard likelihood ratio analysis).

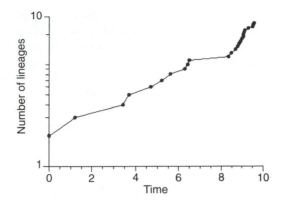

Fig. 11.8 Lineages- through-time plot for salamanders of the genus *Plethodon*.

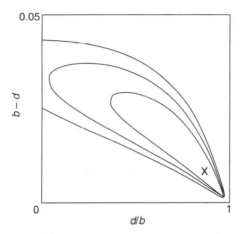

Fig. 11.9 Contour plot of the likelihood surface for the *Plethodon* data. The maximum likelihood estimate, the peak of the surface, is marked by an X.

Rather, we are giving examples of how, if one has accepted a model as reasonable for one's data, one can then use the model to make estimates of parameters of interest. In any particular case, it is the investigators themselves who will decide, on the basis of their interests and knowledge of the group involved, which models are to be considered, compared, and used for inference. Our purpose is to demonstrate that reconstructed phylogenies provide information about more than just the relationships among species.

We have considered several variable rate models elsewhere, including ones in which birth and death change as functions of time or of the number of lineages present (Harvey *et al.* in press *b*; Harvey and Nee in press; Nee *et al.* 1992, 1994). Our analyses of Sibley and Ahlquist's (1990) molecular phylogeny of the birds suggest that, over the period of time from the origin of birds up until the origin of avian families, the overall rate of cladogenesis (defined as lineage birth rate minus lineage death rate) was slowing down: either the lineage birth rate was decreasing through time or the lineage death rate was increasing through time (or both)—possibly as a consequence of a niche-filling process. Subsequent analyses by Robert M. Zink and Joseph Slowinski (pers. comm.) on mitochondrial DNA sequence data suggest that the rate of cladogenesis has also been slowing in nine out of ten avian genera studied. We have also asked whether a reconstructed molecular phylogeny, such as that of the birds, could provide evidence for mass extinction events. The answer is that it could, but even if there was an 80% mass extinction of birds at the end of the Cretaceous, it is unlikely that we should detect clear evidence for it from the molecular phylogeny, particularly if background extinction rates were high relative to speciation rates (Harvey and Nee in press).

11.4 When only a sample of lineages has been analysed

So far, the theory we have been using is appropriate for reconstructed phylogenies that are based on *all* the members of a clade. We now explore the consequences of relaxing this assumption. First, consider the simplest relaxation, and suppose that the reconstructed phylogeny consists of a set of species which have been chosen at random, with respect to their phylogenetic relationships, from a clade which has grown according to a constant rate birth–death process. Figure 11.10 shows that the effect of such sampling is to create the impression of a slowdown in the rate of cladogenesis over time, and that this effect becomes more pronounced the smaller the sample. The effect arises because lineages that have arisen in the recent past are likely to have fewer progeny than lineages which arose in the more distant past and, hence, are less likely to have any progeny represented in the sample.

We can now consider some of the uses which can be made of this theory employing a phylogenetic tree of the retrovirus HIV-1, the 24 sequences isolated primarily from Americans in the mid to late 1980s (Harvey *et al.* in press *a*). It is not a very great leap from macroscopic lineages to viral lineages: the simple birth–death model we are using is a very simple model of the growth of a clade, whether the clade consists of species of macroscopic organisms or lineages of an epidemically spreading disease organism.

Following Harvey *et al.* (in press *a*), Fig. 11.11 shows the increase through time in the actual number of lineages in the molecular phylogeny and Fig. 11.12 shows the theoretically expected increase through time in the number of lineages if the 24 people from whom the virus was isolated represent a very small fraction of the total number of people infected (the fraction 1 in 10 000 was chosen for this figure). The similarity between these two figures gives us some confidence that the theoretical model is not wildly inappropriate, and we will now use it to construct an estimate of the number of people infected. The mathematical theory on which this analysis is based is to be found in Nee *et al.* 1994.

As this is an illustrative example, we will simply assume that R_0 (the number of

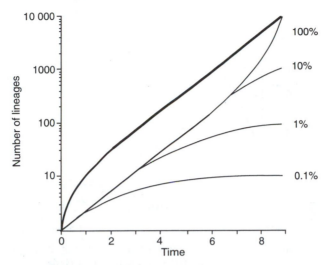

Fig. 11.10 The theoretically expected lineages-through-time plots for reconstructed phylogenies based on successively smaller samples from the actual phylogeny described by the top line.

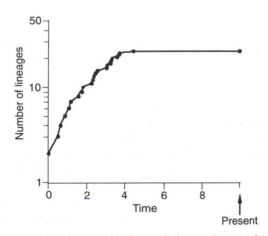

Fig. 11.11 Lineages-through-time plot for a phylogenetic tree of 24 HIV-1 lineages.

new hosts infected by a particular host over the lifetime of the host's infection, Anderson and May 1991) for HIV was 4. In this simple model, R_0 is just b/d. The analysis does not depend on this chosen value, because the estimate of the fraction infected is quite insensitive to R_0 over a reasonable range. A further simplification is achieved by eliminating the parameter $b-d$. This is done by recognizing that, given that we have a sample of 24 lineages, the parameters $b-d$, R_0, f—the fraction in our sample—are not independent. If the parameters $b-d$

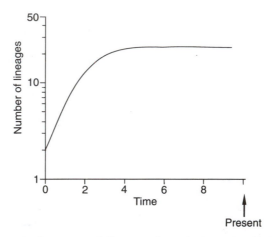

Fig. 11.12 Theoretically expected lineages-through-time plot if 24 taxa (lineages) represent only 1 in 10 000 of the taxa in a clade.

and R_0 are such that millions of people are likely to be infected at the time we sample, then f cannot take large values. We incorporate this constraint in a very simple fashion by defining $b - d$ as a function of f and R_0 such that, using a formula defining the expected total number of people infected, that fraction f will, in fact, give us 24 people.

Figure 11.13 shows the likelihood function for the fraction of infected people represented in our sample. The peak of this surface, the maximum likelihood estimate, is 1 in 10 000, suggesting that, at the time the sample was taken, about a quarter of a million people were infected, which is not completely unreasonable. We emphasize that this analysis is solely to provide an illustrative example: if one was interested in doing this analysis for its own sake, considerably more attention should be paid to detail than we have done here.

The sampling theory has other interpretations. Suppose that we wished to estimate the number of species in a large, little-known clade. If we had a random sample of species in the clade then this could be done using the same analysis, where the fraction, f, is now the fraction of the clade represented in our sample. Such an interpretation raises interesting questions about the relationships between those features of species which are likely to affect our sampling—such as abundance, body size, and range size—and the phylogenetic relationships among the species (Lawton *et al.* 1994). Or suppose that we had reason to believe that a clade had recently suffered a mass extinction (as a result of human activities or otherwise), then, if we had all the extant species in our phylogeny, the same analysis could be used to estimate how many species have gone extinct. Again, interesting questions arise about the interpretation of 'phylogenetically random sampling' when we are discussing extinction. Of course, as we emphasize repeatedly, the particular models we are discussing here do not have any special status: phylogeny interpretation, as a discipline, will consider a large number of

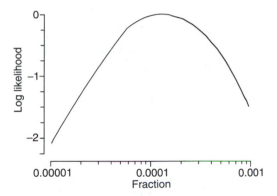

Fig. 11.13   Likelihood function for the fraction of infected people represented in the sample of 24 HIV-1 isolates.

different models. We will now discuss a model which, in some sense, is the complete opposite of the constant birth–death model, and compare their features.

11.5   The constant number model

The constant rate birth–death model describes a clade that is growing in size over time. We may also wish to consider a model in which, although there is speciation and extinction going on, the total number of lineages remains constant through time. This could be a model of density-dependent cladogenesis, for example, in which a 'niche space' is saturated by a group. Such models have already been analysed in the different context of population genetics, where they are known, generically, as the 'coalescent' (Kingman 1982).

We describe the features of relevance here with reference to the Moran model of reproduction, although the results are known to be approximately true for a much wider range of models (Kingman 1982). The Moran model is a model of neutral evolution in a constant-size population with overlapping generations. At each time step, 'generation', one member of the population, chosen at random, reproduces and one member, also chosen at random, dies. We are interested in the distribution of the times between nodes in the genealogy of a sample chosen at random from this population, or of the whole population.

The Moran model is time reversible (Ewens 1979) which means that, if we had a film of the process from which we were calculating statistics, we should calculate the same statistics if we ran the film backwards as forwards. In this sense, we cannot distinguish in which time direction the process is running. This suggests that we can construct a time-reversed Moran model, in which, at each time step, two members of the population are chosen at random and combined to create a new member of the population, and evaluate the same statistics for this model as for the original Moran model.

It so happens that it is much easier to derive the distribution of times between nodes for the time-reversed Moran model (e.g., Watterson 1984) than for the original model. We have a sample of n individuals (lineages) out of a total population of size M. At each time step two lineages of the population are chosen at random and 'coalesced'. What is the probability that these two are both in our sample? The probability that the first one chosen is in the sample is n/M, and the probability that the second one chosen is in the sample, given that the first one is, is $(n-1)/(M-1)$. Hence, the probability that both are in the sample is $n(n-1)/(M(M-1))$. Denoting this probability by $1-u$, we see that the probability that we must wait for n times steps before the 'coalescence' is $(1-u)u^{n-1}$. Hence, the number of steps has a geometric distribution. If $(1-u)$ is small, this can be well approximated by an exponential distribution with parameter $(1-u)$. Hence, the waiting time for the coalescence of n lineages into $n-1$ lineages is approximately exponentially distributed with a parameter proportional to $n(n-1)$. Hey (1992) re-derived the Moran model and suggested its use for the analysis of phylogenies. (Hey actually supposed that the time between steps is exponentially distributed with parameter λ, in which case the waiting time until the coalescence is exactly exponentially distributed, with parameter $\lambda(1-u)$.

So the constant number model leads to exponentially distributed waiting times between the nodes in a phylogeny, such that the parameter of the distribution for the time between the $(n-1)$th and nth lineage is proportional to $(n-1)n$. Figure 11.14 shows the theoretically expected lineages-through-time plot for a sample from a phylogeny which has been constructed by the constant number model. We have the impression of a rapidly accelerating rate of cladogenesis as we approach the present, in stark contrast to the apparent *deceleration* manifested by a sample from a constant birth–death process (Fig. 11.10). We use this comparison to make two points. The first is that appearances can be deceiving: a sample from an expanding clade suggests that the rate of cladogenesis has been decreasing through time, whereas a sample from a clade which is not growing suggests a rapidly increasing rate of cladogenesis! The second point is that different processes leave distinctively different signatures on the reconstructed phylogeny. (In fact, although we shall persist with it here for the sake of continuity, the semilogarithmic representation of the lineages-through-time plot is not the natural one for the constant number model. The approximately hyperbolic growth of the reconstructed phylogeny appears as a straight line if the reciprocal of the number of lineages is plotted against time.)

Whereas the constant rate birth–death model is appropriate for the phylogenies of viruses which are spreading epidemically, the constant number model may be more appropriate for endemic viruses, which infect a roughly constant fraction of the population. To explore this we shall test the fit of this model to hepatitis A virus (HAV). Figure 11.15 shows the lineages-through-time plot for 74 HAV strains from diverse geographical backgrounds. The phylogenetic tree on which the analysis is based was reconstructed on the VP1/2A junction region from those samples considered to be roughly contemporaneous in time (1979–90) using the Fitch–Margoliash algorithm with a molecular clock (PHYLIP

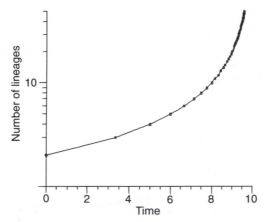

Fig. 11.14 Theoretically expected lineages- through-time plot for a reconstructed phylogeny based on a sample of a clade, if the phylogeny has been generated according to the constant number model described in the text.

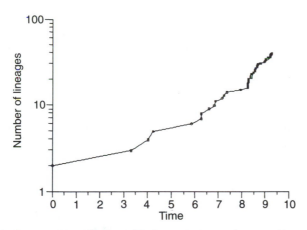

Fig. 11.15 Lineages-through-time plot for a phylogenetic tree of hepatitis A virus.

program KITSCH: Felsenstein 1993). Data are from Robertson *et al.* (1992). Under the hypothesis that this phylogeny really is described by the constant number model, it follows that if the time intervals between the $(n-1)$th and nth lineages are multiplied by $n(n-1)$, then the time intervals are all drawn from an exponential distribution with the same parameter. This hypothesis can now be tested with a χ^2 test (Fig. 11.16), and is consistent with the data (although, we must point out, this is not a particularly powerful test).

Once again, we have used a virus example for illustration. As an aside, we note that it is becoming increasingly clear that conservation biology, with its concern

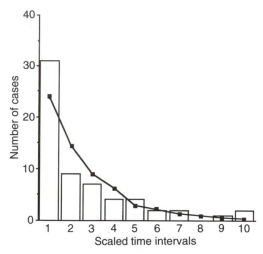

Fig. 11.16   The histogram represents the observed number of scaled time intervals in each category, and the line represents the expected number from a fitted exponential distribution, for the data from Fig. 11.15.

for species' preservation, and epidemiology, with its concern for species' elimination, have vast areas of non-trivial theoretical overlap. We have seen in this Chapter that the same theoretical basis can be used by both disciplines for making inferences from phylogenetic tree structure. Other examples of overlap include the theory of eradication thresholds (Nee 1994; Lawton *et al.* 1994), and the identity of models of the evolution of virulence and the coexistence of competitors in a metapopulation (Nowak and May 1994). We expect this synergistic relationship between epidemiology and conservation biology to deepen over time.

Summary

A molecular phylogeny usually contains no explicit information about rates of lineage extinction, but it does retain information about the evolutionary processes that gave rise to its present structure. Accordingly, molecular phylogenies can be used to reject specific null models of the way we think evolution occurred, including patterns of lineage extinction. Similarly, they can be used to provide maximum likelihood estimates of parameters associated with lineage birth and death rates. Analyses of case studies from vertebrates, invertebrates, and viruses are used to illustrate: (1) how molecular phylogenies provide information about the extent to which particular clades are likely to be under threat from extinction; (2) how cursory analyses of molecular phylogenies can lead to incorrect conclusions about the evolutionary processes that have been at work; and (3) how different evolutionary processes leave distinctive marks on the structure of reconstructed phylogenies.

Acknowledgements

We are grateful to BBSRC (SN), The Wellcome Trust (grant 038468 to PHH), SERC (grant GR/H53655 to PHH), and the Royal Society (RMM) for supporting the work described in this contribution.

References

Anderson, R. M. and May, R. M. (1991). *Infectious diseases of humans*. Oxford University Press.

Caccone, A., Amato, G. D., and Powell, J. R. (1988). Rates and patterns of scnDNA and mrDNA divergence within the *Drosophila melanogaster* subgroup. *Genetics*, **118**, 671–83.

Cox, D. R. and Hinkley, D. V. (1992). *Theoretical statistics*. Chapman & Hall, London.

Cox, D. R. and Lewis, P. A. A. (1966). *The statistical analysis of series of events*. Methuen, London.

Edwards, A. W. F. (1972). *Likelihood*. Cambridge University Press.

Ewens, W. J. (1979). *Mathematical population genetics*. Springer, Berlin.

Felsenstein, J. (1993). PHYLIP (Phylogeny Inference Package), Version 3.5c. Distributed by the Author at Department of Genetics, University of Washington, Seattle WA 98195, USA.

Gould, S. J., Raup, D. M., Sepkowski, J. J., Schopf, T. J. M., and Simberloff, D. S. (1977). The shape of evolution: a comparison of real and random clades. *Paleobiology*, **3**, 23–40.

Harvey, P. H. and Nee, S. Comparing real with expected patterns from molecular phylogenies. *Biol. J. Linn. Soc.*, in press.

Harvey, P. H., Nee, S., Mooers, A. Ø., and Partridge, L. (1991). These hierarchical views of life: phylogenies and metapopulations. In *Genes in ecology*, (ed. R. J. Berry and T. J. Crawford), pp. 123–37. Blackwell, Oxford.

Harvey, P. H., Holmes, E. C., Mooers, A. Ø., and Nee, S. Inferring evolutionary processes from molecular phylogenies. In *Models in phylogeny reconstruction*, Systematics Association Special Volume Series. In press *a*.

Harvey, P. H., May, R. M., and Nee, S. Phylogenies without fossils: estimating lineage birth and death rates. *Evolution*, in press *b*.

Hey, J. (1992). Using phylogenetic trees to study speciation and extinction. *Evolution*, **46**, 627–40.

Highton, R. and Larson, L. A. (1979). The genetic relationships of the salamanders of the genus *Plethodon*. *Syst. Zool.*, **28**, 579–99.

Kingman, J. F. C. (1982). The coalescent. *Stochastic processes and their applications*, **13**, 235–48.

Lawton, J. H., Nee, S., Harvey, P. H., and Letcher, A. J. Animal distributions: patterns and processes. In *Large scale ecology and conservation*, (ed. P. J. Edwards, R. M. May, and N. Webb). pp. 41–58. Blackwell, Oxford 1994.

MacArthur, R. H. and Wilson, E. O. (1967). *The theory of island biogeography*. Princeton University Press, Princeton, NJ.

May, R. M. (1973). *Stability and complexity in model ecosystems*. Princeton University Press, Princeton, NJ.

Nee, S. (1994). How populations persist. *Nature*, **367**, 123–4.

Nee, S., Mooers, A. Ø., and Harvey, P. H. (1992). The tempo and mode of evolution revealed from molecular phylogenies. *Proc. Nat. Acad. Sci. USA*, **89**, 8322–6.

Nee, S., May, R. M., and Harvey, P. H. (1994). The reconstructed evolutionary process. *Phil. Trans. Roy. Soc. Lond.* **B344**, 305–11.

Nowak, M. and May, R. M. (1994). Superinfection and the evolution of parasite virulence. *Proc. Roy. Soc. Lond. B*, **255**, 81–89.

Raup, D. M., Gould, S. J., Schopf, T. J. M., and Simberloff, D. S. (1973). Stochastic models of phylogeny and the evolution of diversity. *J. Geol.*, **81**, 525–42.

Robertson, B. H. *et al.* (1992). Genetic relatedness of Hepatitis A virus strains recovered from different geographical regions. *J. Gen. Virol.*, **73**, 1365–77.

Sibley, C. G. and Ahlquist, J. E. (1990). *Phylogeny and classification of birds.* Yale University Press, New Haven.

Slowinski, J. B. and Guyer, C. (1993). Testing whether certain traits have caused amplified evolution. *Amer. Nat.*, **142**, 1019–24.

Stanley, S. M. (1979). *Macroevolution: Pattern and process.* W. H. Freeman, San Francisco.

Thompson, E. A. (1975). *Human evolutionary trees.* Cambridge University Press.

Watterson, G. A. (1984). Lines of descent and the coalescent. *Theor. Pop. Biol.*, **26**, 77–92.

12

Biological models for monitoring species decline: the construction and use of data bases

C. R. Margules and M.P. Austin

12.1 Introduction

A necessary prerequisite for estimating extinction rates is a database that adequately represents the biota and is comprehensive in its geographical coverage. This is a fundamental and obvious requirement. So obvious, in fact, that it is often taken for granted that such data exist. In reality, only a few small parts of the world with a long history of biological field surveys, such as the British Isles, can lay claim to adequate and comprehensive databases and even these can only be considered adequate for some components of the biota at some scales. Usually, field records are collected in a haphazard or opportunistic manner; the species recorded are the ones of interest to the collector; and the places they are recorded from are the places those species might be expected to be found, or are conveniently accessible. Accordingly, extensive and often detailed collections of field records in museums, herbaria, and various natural resource management agencies throughout the world are flawed, because they are incomplete and often biased, both in geographical coverage and in the sense that they are records of subsets of taxa.

The purpose of this chapter is to describe some recent advances in the design of biological surveys and the analysis of survey data to map more accurately the spatial distribution patterns of species and thus improve the baseline against which extinction rates can be estimated.

There are four requirements for establishing a sound ecological database. These are: (1) a conceptual framework based on ecological theory; (2) field survey design principles explicitly based on the conceptual framework; (3) a rationale for determining which measurements should be made at the chosen field sample sites in addition to records of the target species; and (4) appropriate statistical methods for analysing survey data and predicting wider distribution and/or abundance patterns from the point records that the field sample sites represent.

For confident predictions of extinction probability in the wild, ecological theory and methods will have to be developed to link the patterns in distribution and abundance estimated via steps 1–4 above, to population parameters.

Each of these four requirements is considered in turn below.

KOALA SIGHTINGS

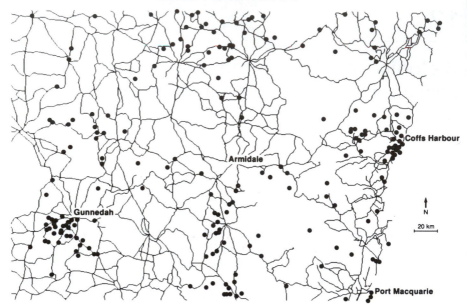

Fig. 12.1   Koala records and the road network on part of the New South Wales north coast. (Courtesy of New South Wales National Parks Wildlife Service.)

1   *The conceptual framework*

Figure 12.1 is a map of koala sightings from part of the mid-north coast of New South Wales, Australia. Notice that these records closely map the road network. They are *ad hoc* and opportunistic. It is not possible, from this map, to define the range of this species. The sites with records are almost certainly a subset of the sites this species actually occurs on and there are few, if any, records of where it was looked for but not found; that is, sites with recorded absences. The koala is probably Australia's most charismatic animal but it is still not possible to define its range even in an area close to major population centres.

This is only one example of a pervasive problem with most existing data bases. Field records of trees in the Amazon, for instance, map the river network (G. Prance, pers. comm.). Such databases contain recorded locations of some species but not all (or most) actual locations and no records of absences. They are not even representative of, let alone sample adequately, real distribution patterns.

The distribution and abundance patterns of species are not random or uniform. Plant community ecologists have adopted the concept of the individualistic continuum to explain observed patterns of variation in vegetation (Gleason 1926). This concept holds that each species has a unique distribution determined by its genetic make-up and physiological requirements, which is constrained by ecological interactions with other species. It is closely related to the concept of the niche as often used by animal ecologists (Whittaker *et al.* 1973; Austin 1985) and similar continuum patterns apply to animals (Rotenberry and Wiens 1980).

The continuum concept implies that: (a) each species occupies its own niche not easily predicted from that of other species; (b) species distribution patterns are most accurately defined in a multidimensional environmental space; (c) the resultant spatial pattern shows high robust populations in scattered locations with preferred habitat and lower, less secure populations in areas of more marginal habitat; and (d) surveys which sample pre-existing mapping units, such as vegetation communities or formations, are unlikely to adequately estimate population patterns.

This is an appropriate conceptual framework for ecological survey design because it links species distribution patterns with variation in the environment. Whittaker (1956, 1960) and Perring (1958, 1959, 1960) provide a rationale for selecting the types of environmental gradients that should be incorporated in survey designs. More recently, Nix (1982) argued that, for the purpose of estimating distribution patterns of plants and animals, complete niche specification is not necessary and that the five environmental regimes, namely, solar radiation, temperature, moisture, mineral nutrients, and other components of the biota, are sufficient.

The goal of an ecological survey is accurately to detect species distribution patterns in both environmental and geographic space. The lesson for survey design is that the best estimates of which species occur in a region, and the patterns of abundance and range they exhibit, require the region to be stratified using major resource gradients, or environmental variables, such as temperature, moisture, and substrate, and to ensure that the range of combinations of these variables is sampled. While the major determinants of patterns of occurrence are environmental, distributions are also conditional on co-occurring species. Thus, it is necessary to record a range of species from survey sites in order to examine processes influencing persistence or extinction. Biological collections alone are insufficient for this purpose. Ecological databases are needed.

2   Field survey design

Survey data should be accurate and reliable and fairly represent the true distribution and abundance patterns of the species recorded. Yet the design of surveys is a neglected topic (Austin and Heyligers 1991). Surveys themselves can be tedious, time-consuming, and labour-intensive. Biological survey is not seen as a scientific endeavour and is therefore ignored by textbooks. There now exists a very sophisticated and still rapidly evolving technology for displaying and manipulating data in computers but the methodology for acquiring those data in the first place remains primitive. The design of a survey has such profound implications for the subsequent use of the data that rigorous design rules should be formulated, explicated, and applied.

Environmental stratification is an appropriate conceptual framework, but devising efficient and effective surveys, even within this framework, is an immensely practical problem. Strict statistical sampling methods can be prohibitively expensive. Gillison and Brewer (1985) proposed the use of gradient directed transects, or gradsects, as a practical tool for designing surveys efficiently. The idea is to identify a set of transects which intercept the major environmental strata. If these transects (gradsects) are aligned along gradients of steep environmental change, then the greatest amount of environmental change can be intercepted in the shortest distance, maximizing the cost-effectiveness of the survey. Austin and Heyligers (1989) proposed refinements of this idea, including replication within transects and explicit rules for locating field sample sites, which incorporate another lower level of stratification.

An example: trees in a coastal hardwood forest, NSW, Australia

Austin and Heyligers (1989, 1991) conducted a survey of tree species in the *Eucalyptus* forests of coastal northern New South Wales. The area to be surveyed covered 20 000 km^2, the entire catchments of five major rivers. These rivers originate on tablelands and flow east through deeply dissected mountains and hills on to a coastal plain and then into the Tasman Sea. Most of the area remains forested, though significant parts of the broader valleys and the coastal plain have been cleared for agriculture and settlement.

Austin and Heyligers' design protocol contained 7 steps:

(1) identify the major environmental variables influencing the distribution patterns of vegetation;

(2) recognize the subset of environmental variables best suited to determining the position and direction of gradsects so that they sample the range of combinations of environmental variables;

(3) choose the best available data for environmental stratification and the best available technology for implementing gradsect selection;

(4) stratify the environment within gradsects and break the gradsects up into segments to allow replicate sampling at different geographical locations;

(5) decide whether or not another level of stratification is needed to take account of environmental variation at the local scale;

(6) decide on the effort that should be spent sampling the rarest environmental strata as opposed to increasing replication of the common strata; and

(7) be flexible because some sample sites selected in the laboratory will not be available in the field if, for example, they have been cleared or access is denied. New sites have to be chosen following established rules.

Survey design   Austin *et al.* (1984) had shown previously that rock type, rainfall, and temperature have a strong influence on the distribution of tree species at this regional scale. As temperature correlates strongly with altitude, altitude was used because it could be easily determined in the field.

A regular 0.01 degree grid (*C.* 1 km^2) was placed over the study area. Possible gradsect locations were evaluated by plotting all grid points against rainfall, altitude, and rock type. These three major variables were divided into classes; 9 for rock type, 7 for altitude, and 8 for rainfall, producing $9 \times 7 \times 8 = 504$ possible

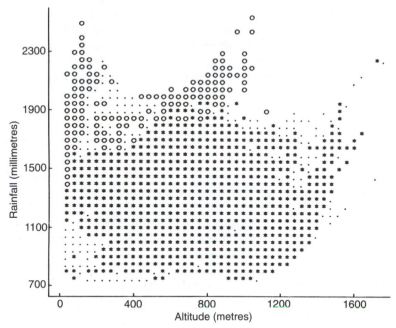

Fig. 12.2 Combinations of rainfall and altitude in the north coast survey area. *, combinations represented in three gradsects; ○, new combinations represented by the addition of the fourth gradsect; •, combinations not represented by any gradsect. (From Austin and Heyligers 1989, reproduced with permission of Elsevier Science Publishers.)

combinations or environmental cells. In fact only 215 occur within the study area, mainly due to the localized distribution of rock types. Four gradsects were chosen which together most adequately represented the rainfall/altitude/rock type combinations of the study area. Figure 12.2 helps to explain this process. The four gradsects illustrate the inevitable trade-off between completeness of coverage and resources available for the survey. Four gradsects was the maximum number that could be sampled with available resources and those four were positioned to encompass the maximum number of rainfall/altitude/rock type combinations. Of the 43 combinations not covered, 18, nearly half, were represented in the study area by only one or two grid points. Figure 12.3 shows the selected gradsects, the distribution of sample points within them, and records of one species, which is used to exemplify analytical techniques below.

Sampling strategy Ideally, each grid point in each environmental cell should have an equal, or at least known, probability of being sampled. In practice, this is impossible because access to many grid points would impose unacceptable costs (e.g., use of helicopters). Instead, Austin and Heyligers adopted a set of rules which ensured consistency of sampling and, being explicit, provide the opportunity for the degree of bias to be determined.

The rules included restricting sampling to within 0.5 km of access tracks and proportional sampling depending on the number of grid points per environmental cell. Sample sites were selected randomly within each geographic segment. Each sample site consisted of five quadrats which together cover local environmental variation due to aspect and topographic position. Further details of quadrat location and field measurements can be found in Austin and Heyligers (1989, 1991).

3   Field measurements

The rationale for deciding on what to measure at each field sample site is determined by the requirement to predict wider geographic range and abundance patterns from the samples. Thus, field measurements, aside from records of the target species, should be made of those variables most likely to correlate with species distribution patterns.

Austin *et al.* (1984, 1990). Margules and Nicholls (1987), Margules and Stein (1989), and Nicholls (1989, 1991) have successfully used environmental variables, such as rainfall, temperature, lithological substrate, and solar radiation, to model the distribution patterns of plant species and plant communities. Braithwaite *et al.* (1984, 1989) and McKenzie *et al.* (1989) provide some empirical support for a similar relationship between environmental variables and some animal distributions at a regional scale.

Ecological knowledge and experience have to be brought to bear. The most appropriate variables may differ in different ecosystems or biomes but the rationale is that they should be expected to be adequate predictors of wider distribution patterns. Austin and Heyligers used altitude, aspect, slope, and topographic position in addition to rainfall and temperature estimates and rock type. These seven variables were deemed to be appropriate and sufficient correlates of wider distribution patterns for trees in the forests of their study area, on the basis of previous experience and local knowledge.

4   Analytical methods

An appropriate analytical technique for survey data collected in an explicit systematic way is generalized linear modelling (GLM) (McCullagh and Nelder 1983), which has traditional least-squares regression as a special case, is flexible in its assumptions and allows the simultaneous use of continuous variables and factors. It is a tool that allows prediction from point samples via correlation with external variables to the wider geographic space. If the survey is complete, representing a true sample of the environmental space, the predictions will be interpolations with a high degree of confidence. If there are combinations of environmental variables not sampled in the survey then the predictions will be extrapolations beyond the domain of the data and there will be less confidence in their accuracy.

Nicholls (1989) modelled the distribution of one tree species from Austin and Heyligers' survey data, *Eucalyptus radiata*, which is reported here as an example. Other examples using different data sets can be found in Austin *et al.* (1984, 1990), Margules *et al.* (1987), Margules and Nicholls (1987), and Margules and Stein

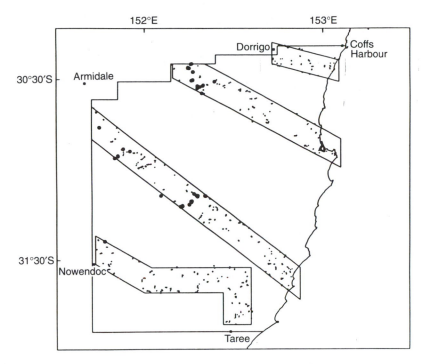

Fig. 12.3   The north coast survey area showing the location of gradsects. The large dots represent survey sites where *Eucalyptus radiata* was found and the smaller dots all other survey sites. (From Nicholls 1989, reproduced with permission from Elsevier Science Publishers.)

(1989). Leathwick and Mitchell (1992) provide a New Zealand example based on a conceptual framework which incorporates historical disturbance following volcanic activity.

Figure 12.3 shows the recorded geographical distribution of *Eucalyptus radiata* and Fig. 12.4 shows its environmental distribution in the space defined by rainfall and temperature. Table 12.1 lists the environmental variables available to Nicholls for use as predictors. Three of them are continuous and three are categorical.

Nicholls used the presence or absence of the species at a survey sample site as the dependent variable, and hence logistic regression. He adopted a forward stepwise procedure, appropriate for an exploratory analysis of this kind. Complete details of the fitting procedure plus the parameter estimates and standard errors of the final model are supplied by Nicholls (1989).

The probability of *Eucalyptus radiata* occurring at all combinations of rainfall, temperature, and rock type which are found in the study area was calculated and the resulting surfaces contoured (e.g., Fig. 12.5). A geographic map of the probability of occurrence of *E. radiata* was generated by calculating the probability of occurrence in each $\frac{1}{100}$th degree grid cell using rainfall, temperature, and the

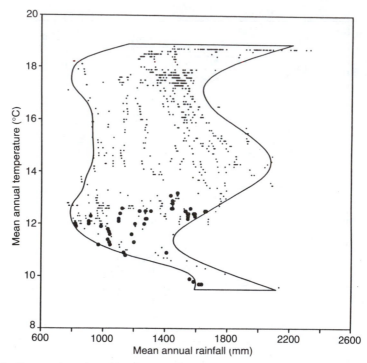

Fig. 12.4   Survey sites plotted in the space formed by mean annual temperature and mean annual rainfall. The large dots represent sites where *Eucalyptus radiata* was recorded. (From Nicholls 1989, reproduced with permission of Elsevier Science Publishers.)

Table 12.1   Details of the environmental variables available for inclusion in model developed to predict the probability of occurrence of *Eucalyptus radiata*. (From Nicholls 1989.)

Variable	Type	Range or no. of levels
Altitude	Continuous	0–1750 m
Mean annual rainfall	Continuous	800–2300 mm
Mean annual temperature	Continuous	9.0–19.9 °C
Lithology	Categorical	9
Topography	Categorical	6
Exposure	Categorical	3

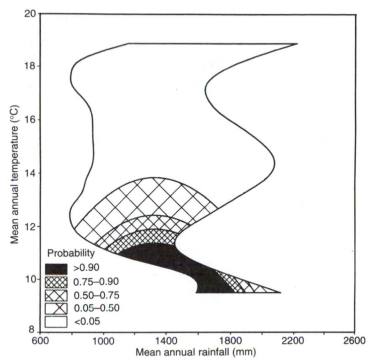

Fig. 12.5 Contours of the predicted probability of *Eucalyptus radiata* occurring in the rainfall/temperature space on one rock type, coarse-grained sediments, from the model in Table 12.1. (From Nicholls 1989, reproduced with permission from Elsevier Science Publishers.)

appropriate rock type. Figure 12.6 shows that map, which can be compared with Fig. 12.3.

12.2   Development of theory

The practical problem managers face is that of monitoring and predicting the probability of extinction in the wild, that is, in landscapes with multiple species in heterogeneous populations and spatially and temporally variable environments. Much of the traditional theory of population dynamics applies to homogeneous populations and does not adequately incorporate environmental variation. Thus, estimates of extinction risk for most practical applications will require newer theories dealing with small or declining populations, with metapopulation or nuclear structure, in fluctuating or trending environments (e.g., Gilpin and Hanski 1991; Caughley 1994; Lawton, Chapter 10).

The procedures outlined in this chapter are for deriving a regional ecological database showing the estimated distribution patterns of species. If quantitative data are available, then estimates of population size can also be obtained.

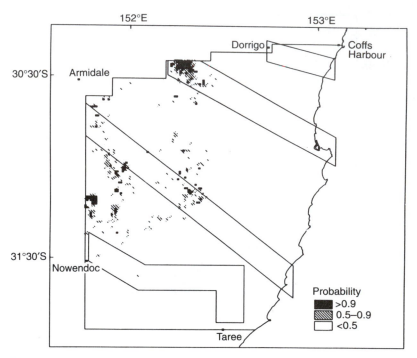

Fig. 12.6 Predicted geographic distribution of *Eucalyptus radiata*, which can be compared with the recorded distribution within gradsects shown in Fig. 12.3. (From Nicholls 1989, reproduced with permission of Elsevier Science Publishers.)

Population size alone, however, does not usually provide an estimate of population viability or probability of extinction. The newer theories mentioned above will, in turn, have to be linked to databases containing distribution patterns in both environmental and geographic space, themselves based on additional theory which relates the position of a population in environmental space to the demographic and genetic processes determining population viability.

In the short term the greatest urgency, and therefore the bulk of research and management resources, is directed at species in small, relatively homogeneous populations because these are the species at greatest current risk. In the medium to long term, however, there will be no substitute for adequate databases linked to appropriate population dynamics theory for the management of biological diversity in the wild.

12.3 Discussion

Better databases can be compiled with systematic stratified surveys based on a sound conceptual framework and the use of relevant analytical techniques. All of the methods described here are published and the software is widely available. The ideas behind them are essentially common sense. They were developed from the premise that complete inventories of species and populations within regions or biomes are not a realistic option in the foreseeable future, if ever.

Collecting expeditions have not been conducted in the past with a view to mapping range and abundance patterns of species. Yet that is what collections in museums and herbaria are now being used for because it is range and abundance patterns that need to be known if extinction rates are to be monitored accurately. Future collecting activities should incorporate environmental stratification in an explicit survey design. Green and Gunawardena (1993) have surveyed both the flora and fauna of Sri Lankan forests using the gradsect design approach.

When the expense and labour invested in traditional collecting expeditions is considered, designed surveys will probably prove to be cheaper because they sample the same area more efficiently. Survey costs are almost never published (although see Margules and Austin 1991). Cost-effective surveys require both presence and absence data for predictive modelling (Margules and Austin 1991). However, there are empirical methods for estimating distribution patterns of species from presence-only data such as museum records. Perhaps the best known is BIOCLIM (Nix 1986; Busby 1991) which compares the climatic profile of sites from which species have been recorded, with other sites to locate similar climatic profiles and therefore potential locations of those species. This is a rational way to use presence-only data, but the predictions are not attended by the same confidence that they would be if the modelling technique was statistical. Museum records are also being used now in an innovative way to help identify global-priority areas for conservation action based on taxonomic diversity (Vane-Wright et al. 1991; Pressey et al. 1993).

Nevertheless, these are not reasons to continue with ad hoc or opportunistic collections of biological records. If future collections and other surveys were to be carefully planned based on environmental stratification and explicit design rules, the data could be used more effectively and with more confidence to estimate wider distribution patterns at a regional scale. One practical use of the resultant database would be to identify a subset of sites within a region that together sample, in a statistical sense, the biological diversity of that region as it is expressed in the database (Margules and Nicholls 1994). This subset can be thought of as a nominal reserve network which samples regional diversity and is therefore a suitable basis for developing regional conservation plans (Margules et al. in press).

Although ecological surveys need a conceptual framework (Austin and Smith 1989; Austin and Heyligers 1989) they must also be informed by experience, intuition, common sense, and local ecological knowledge. Ecological and evolutionary history play a role, in some cases a major role, in determining distribution patterns. Species with relict distributions or vicariant species occupying similar, but geographically isolated, environments, will be detected with geographic replication but may not be modelled adequately. Plant species in the hyper-diverse areas of the Cape region, South Africa (Cowling 1992) and south-western Western Australia may be cases in point.

Nevertheless, databases derived from explicitly designed ecological surveys and statistical modelling using the idea of an individualistic continuum as the conceptual framework is a first descriptive step towards understanding population dynamics in space and time, and, therefore, estimating more accurately probabilities of extinction.

Summary

We describe procedures for designing ecological surveys and analysing survey data, which more accurately estimate the spatial distribution patterns of species than the opportunistic or *ad hoc* biological field collections generally available. Surveys should be based on an environmental stratification and the data collected should be sufficient to enable statistical estimates of wider spatial distribution patterns to be made from the point sample sites of the survey itself. Without ecological databases that map spatial distribution patterns, accurate estimates of extinction rates will remain problematical; that is, confined to single populations and geographically restricted.

Acknowledgements

Yrjö Haila and Albert van Jaarsveld commented critically on a draft of the manuscript, which was also improved considerably by comments from Robert May and John Lawton. Paul Walker made the koala data available and Heather Lynch produced the map (Fig. 12.1). Jacqui Meyers drew the other figures.

References

Austin, M. P. (1985). Continuum concept, ordination methods and niche theory. *Ann. Rev. Ecol. Syst.*, **16**, 39–61.

Austin, M. P., and Heyligers, P. C. (1989). Vegetation survey design for conservation: gradsect sampling of forests in north-eastern New South Wales. *Biol. Conserv.*, **50**, 13–32.

Austin, M. P., and Heyligers, P. C. (1991). New approach to vegetation survey design: gradsect sampling. In *Nature conservation: Cost effective biological surveys and data analysis*, (ed. C. R. Margules and M. P. Austin), pp. 31–6. CSIRO, Melbourne.

Austin, M. P., and Smith, T. M. (1989). A new model for the continuum concept. *Vegetatio*, **83**, 35–47.

Austin, M. P., Cunningham, R. B., and Fleming, P. M. (1984). New approaches to direct gradient analysis using environmental scalars and statistical curve-fitting procedures. *Vegetatio*, **55**, 11–27.

Austin, M. P., Nicholls, A. O., and Margules, C. R. (1990). Measurement of the realized qualitative niche: environmental niches of five *Eucalyptus* species. *Ecol. Monogr.*, **60**, 161–77.

Braithwaite, L. W., Turner, J., and Kelly, J. (1984). Studies of the arboreal marsupial fauna of eucalypt forests being harvested for woodpulp at Eden, New South Wales. III. Relationships between fauna densities, eucalypt occurrence and foliage nutrients and soil parent materials. *Austr. Wildlife Res.*, **11**, 41–8.

Braithwaite, L. W., Austin, M. P., Clayton, M., Turner, J., and Nicholls, A. O. (1989). On predicting the presence of birds in *Eucalyptus* forest types. *Biol. Conserv.*, **50**, 33–50.

Busby, J. R. (1991). BIOCLIM-a bioclimatic analysis and prediction system. In *Nature conservation: Cost effective biological surveys and data analysis*, (ed. C. R. Margules and M. P. Austin). pp. 64–8. CSIRO, Melbourne.

Caughley, G. (1994). Directions in conservation biology. *J. Anim. Ecol.*, **63**, 215–24.

Cowling, R. M. (ed.) (1992). *The ecology of fynbos*. Oxford University Press.

Gillison, A. N., and Brewer, K. R. W. (1985). The use of gradient directed transects or gradsects in natural resource survey. *J. Environ. Man.*, **20**, 103–27.

Gilpin, M., and Hanski, I. (ed.) (1991). *Metapopulation dynamics: Empirical and theoretical investigations*. Academic Press, London.

Gleason, H. A. (1926). The individualistic concept of the plant association. *Bull. Torrey Bot. Club.*, **53**, 1–20.

Green, M. J. B., and Gunawardena, E. R. N. (1993). *Conservation evaluation of some natural forests in Sri Lanka*. Forestry Department, Sri Lanka, in association with UNDP, FAO, and IUCN, Sri Lanka.

Leathwick, J. R., and Mitchell, N. D. (1992). Forest pattern, climate and vulcanism in central North Island, New Zealand. *J. Veg. Sci.*, **3**, 603–16.

Margules, C. R., and Austin, M. P. (ed.) (1991). *Nature conservation: Cost effective biological surveys and data analysis*. CSIRO, Melbourne.

Margules, C. R., and Nicholls, A. O. (1987). Assessing the conservation value of remnant habitat 'islands': mallee patches on the western Eyre Peninsula, South Australia. In *Nature conservation: the role of remnants of native vegetation*, (ed. D. A. Saunders, G. W. Arnold, A. A. Burbidge, and A. J. M. Hopkins), pp. 89–102. Surrey Beatty & Sons in association with CSIRO and CALM, Chipping Norton, New South Wales.

Margules, C. R., Nicholls, A. O., and Austin, M. P. (1987). Diversity of *Eucalyptus* species predicted by a multi-variable environmental gradient. *Oecologia*, **71**, 229–32.

Margules, C. R., and Nicholls, A. O. (1994). Where should nature reserves be located? In *Conservation biology in Australia and Oceania*, (ed. C. Moritz, J. Kikkawa, and D. Doley), pp. 10–25. Surrey Beatty & Sons, Chipping Norton, New South Wales.

Margules, C. R., and Stein, J. L. (1989). Patterns in the distributions of species and the selection of nature reserves: an example from *Eucalyptus* forests in south-eastern New South Wales. *Biol. Conserv.*, **50**, 219–38.

Margules, C. R., Cresswell, I. D., and Nicholls, A. O. A scientific basis for establishing networks of protected areas. In *Systematics and conservation evaluation*, (ed. P. L. Forey, C. J. Humphries, and R. I. Vane-Wright). Oxford University Press. in press.

McCullagh, P., and Nelder, J. A. (1983). *Generalised linear models*. Chapman & Hall, London.

McKenzie, N. L., Belbin, L., Margules, C. R., and Keighery, G. J. (1989). Selecting representative reserve systems in remote areas: a case study in the Nullarbor region, Australia. *Biol. Conserv.*, **50**, 239–61.

Nicholls, A. O. (1989). How to make biological surveys go further with generalised linear models. *Biol. Conserv.*, **50**, 51–75.

Nicholls, A. O. (1991). Examples of the use of generalised linear models in analysis of survey data for conservation evaluation. In *Nature conservation: Cost effective biological surveys and data analysis*, (ed. C. R. Margules and M. P. Austin), pp. 54–63. CSIRO, Melbourne.

Nix, H. A. (1982). Environmental determinants of biogeography and evolution in Terra Australia. In *Evolution of the flora and fauna of arid Australia*, (ed. W. R. Baker and P. J. M. Greenslade), pp. 47–66. Peacock, Adelaide, South Australia.

Nix, H. A. (1986). A biogeographic analysis of Australian Elapid snakes. In *Atlas of Elapid snakes of Australia*, Australian Flora and Fauna Series, No. 7, (ed. R. Longmore), pp. 4–15. Australian Government Publishing Service, Canberra.

Perring, F. (1958). A theoretical approach to a study of chalk grassland. *J. Ecol.*, **46**, 665–79.

Perring, F. (1959). Topographical gradients in chalk grassland. *J. Ecol.*, **47**, 447–81.

Perring, F. (1960). Climatic gradients in chalk grassland. *J. Ecol.*, **48**, 415–42.

Pressey, R. L., Humphries, C. J., Margules, C. R., Vane-Wright, R. I., and Williams, P. H. (1993). Beyond opportunism: key principles for systematic reserve selection. *Trends Ecol. Evo.*, **8**, 124–8.

Rotenburg, J. T. and Weins, J. A. (1980). Habitat structure, patchiness, and avian communities in North American steppe vegetation: a multivariate analysis. *Ecology*, **6**, 1228–50.

Whittaker, R. H. (1956). Vegetation of the Great Smoky Mountains. *Ecol. Monogr.*, **26**, 1–80.

Whittaker, R. H. (1960). Vegetation of the Siskiyou Mountains, Oregon and California. *Ecol. Monog.*, **30**, 279–338.

Whittaker, R. H., Levin, S. A., and Root, R. B. (1973). Niche, habitat and ecotope. *Amer. Nat.*, **107**, 321–38.

Vane-Wright, R. I., Humphries, C. J., and Williams, P. H. (1991). What to protect? Systematics and the agony of choice. *Biol. Conserv.*, **55**, 235–54.

13

Classification of threatened species and its role in conservation planning

Georgina M. Mace

13.1 Introduction

Over the next several decades, actions will need to be taken towards preserving the many species now facing extinction. Unfortunately, this will be done in the context of enormous ignorance about most of those species, with limited resources, and with continuing developments that contribute to species' extinction. In this chapter, I review the role that categorizing species according to their perceived risk of extinction can play, and outline some of the practical advantages and disadvantages of this approach. In particular, I shall focus on the fact that categorizing risk levels is only a first step. Deciding what to do next and developing rational methods for setting priorities will be just as important, and presents additional challenges.

13.2 The role of threatened species lists

Threatened species lists are produced most commonly in the Red Lists and Red Data Books of IUCN—the World Conservation Union. These were formally established in the early 1960s (Scott *et al.* 1987*b*) and although their size, format, and style has evolved since that time, the basic concept of providing readily assimilated information to focus attention on the plight of endangered species remains. The Red Data Book concept has been very successful, and there are now many regional, national, and taxonomic lists based on it (Burton 1984; R. Fitter and M. Fitter 1987; Thomas and Morris, Chapter 8).

Early on it was clear that general awareness-raising was most important, and species were selected partly on the basis of their appeal (Talbot 1959). However, by the mid-1960s there was a more serious attempt to make the lists comprehensive, although it soon became clear that this was going to be impossible for many plants and invertebrates (Scott *et al.* 1987*b*).

Increasingly, the lists have been used for more than just raising awareness and have been applied to setting priorities for species conservation. In this context it is important that the process for categorizing species is objective, standardized, and equally applicable across broad taxonomic groups whose basic biology and life histories differ fundamentally. The current categorization system is perceived to have problems in this regard, and recently, steps have been taken towards revising methods used to categorize species. (R. Fitter and M. Fitter 1987; Mace and Lande 1991; Master 1991; Mace et al. 1992).

13.3   Problems with threatened species categories

Apart from these difficulties over methodology, there are other general criticisms of threatened species lists, which also need to be taken into account in reviewing their role. First is the concern that a focus on species alone will be insufficient to meet the general needs of biodiversity conservation (Scott et al. 1987a). There has recently been much progress in the development of analytical methods to set priorities among areas for conservation on the basis of some measure of their species complement (Forey et al. in press). However, species are the units of extinction considered here, and an understanding of their dynamics is crucial to a consideration of community and ecosystem processes, and to the development of effective conservation strategies for all of these (Pimm 1992). Recent empirical studies have shown that where species data are poor, area-based methods that select for diversity can be effective if a sufficient number of sites can be selected, but incorporation of species information provides a much better selection (Thomas and Mallorie 1985). In addition, where species play a 'flagship' role, they can provide an effective method for ecosystem preservation (Dietz et al. 1994; Losos 1993). Species-based approaches should be a component of area-based methods, and the two should be used in conjunction and not as alternatives.

A second problem is that published lists of threatened species inevitably focus on well-known forms. Two large sets of species, the undescribed and the unstudied are, to their detriment, under-represented (Diamond 1989). Effective species conservation involves at least three different activities: describing species, researching their status, and protecting them. Threatened species classifications only fulfil a part of this process and should not eclipse the others. The system should not prejudice the poorly known species, but equally should be rigorous in the analysis of any existing species data (see below). This is one of the most challenging areas in developing systems for categorizing threat.

A major problem with existing Red Lists is that, for the most part, they have not been comprehensive in their review of species in particular taxonomic groups or geographic regions. Species are generally listed when a biologist becomes aware that a species has a problem; they do not often result from any systematic review (Diamond 1989). As WCMC (1992) note, among the higher taxa in the IUCN (1990) Red List, only the birds had been fully considered. About 50% of mammal species, and probably less than 20% of reptiles, 10% of amphibians, 5% of fish, and a much smaller proportion of invertebrates had been reviewed. Therefore, non-inclusion in a Red List can mean either that a species has been reviewed and its status considered sufficiently secure, or that the species, for whatever reason (usually lack of information), was not considered. Clearly, these

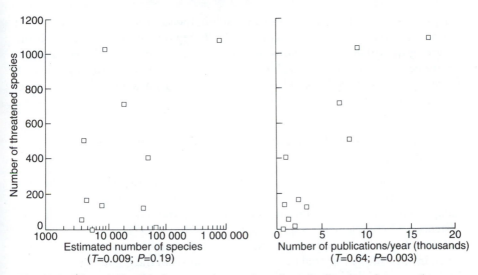

Fig. 13.1 The relationship between the number of species listed as threatened among major taxa (mostly phyla) (WCMC 1992) and the estimated total species number (left-hand graph), and the number of papers published on the same groups (from May 1988) (right-hand graph). The taxa are mammals, birds, reptiles, amphibians, fish, molluscs, annelids, insects, arachnids, crustaceans, and echinoderms.

two need to be distinguished explicitly. One solution is to regard all species as endangered until proven otherwise—the so-called Green List (Imboden 1987; Diamond 1987). While this suggestion has a certain logic and attraction, it is probably unworkable and might detract attention away from the task of collecting information about the poorly known species.

Nevertheless, it is clear that the highest proportion of threatened species are found among the well-known birds and mammals (WCMC 1992). Intensity of interest is hard to measure, but one available index is the number of papers published on different major taxa. In Fig. 13.1, the numbers of published papers cited in *Zoological Record* (from May 1988) is significantly and positively correlated with the number of species listed as threatened in the same groups in the IUCN 1990 Red List (from WCMC 1992). There is no relationship between the estimated number of described species and the number listed as threatened. There are, of course, other possible explanations for this pattern. In particular, large-bodied forms may be more extinction-prone than small ones (Diamond 1984; Belovsky 1987, but see Lawton, Chapter 10) and for practical reasons less attention may be focused on small-bodied species. However, this is unlikely to be a complete explanation (WCMC 1992, Smith *et al.* 1993*b*).

A final problem is that in both major systems used today, the IUCN (1990) definitions and the United States Endangered Species Act (see Rohlf 1989), the definitions for endangered status are merely stated in terms of '. . . in danger of extinction . . .'. Without any reference to a time-frame, or to the likelihood of extinction within this time-frame, the categories may be interpreted in different ways by different authorities. This can result in the same species being classified quite differently, or in different taxa or different regional floras and faunas being judged by quite different standards. If conservation action is to be based on them, threatened species categories will require an approach that is better validated. A more objective system that can be challenged and judged against an accepted set of rules is needed.

13.4   Systems for classifying species according to risk of extinction

With this background to the role of threatened species categories, a review of a range of recently published systems is presented in Table 13.1. It is not comprehensive; rather it aims to provide a representative sample of recent systems, and complements a full review by Munton (1987).

Most of these are based around existing IUCN definitions, usually with some amendments. Many regional threatened species lists in the United States use the Endangered Species Act definitions as a basis (see also Munton 1987). In addition, the classification of rarity by Rabinowitz (1981) has been used and two more recent proposals based on quantitative criteria (Mace and Lande 1991, Mace et al. 1992) have been adopted for some well-studied taxonomic groups. Others have been developed entirely independently (Table 13.1).

The main conclusion from the information presented in Table 13.1 is that threatened species listings measure a number of characteristics that are not directly related to the extinction risk. Sometimes this is done overtly by separating these into distinct categories. For example, 'out of danger' in the IUCN system does not qualify a species for threatened status. But many characteristics that could lead to listing in a threatened category do not necessarily reflect threat. For example, a restricted geographic distribution is sufficient to qualify a species for inclusion in the IUCN category 'rare', and the category 'insufficiently known' includes species that are 'suspected' to belong in one of the threatened categories, but about which there is little information. These categories are therefore measuring multiple characteristics of species, many of which may have a bearing on planning for their conservation, but not necessarily reflecting their risk of extinction.

All systems reviewed use small population size specifically in defining threatened status, though only a minority (5/19) present any quantitative guidelines. The effect of turning from qualitative to quantitative definitions may be significant. In less well-known groups it is likely that the number in the threatened category will decline, because of the difficulties of applying quantitative criteria to poorly known species. However, in general, the application of quantitative criteria to well-studied vertebrate groups has led to an increase in the number of species listed as threatened (Seal et al. 1993; Mace 1994). Some consequences are exemplified in the categorizations made by Green (1992) for waterfowl and Osborne (in press) for cycads. Both these groups had previously been classified by the IUCN criteria. In both cases there was an

[1] Taxon	[2] Region	[3] Based on	[4] In danger of extinction	[5] Will become endangered	[6] Pop. size small	[7] Small range or no. of sites	[8] Commercial or specific threats	[9] Recovered from threat	[10] Protection status	[11] Declining	[12] Declined but not threatened	[13] Suspected threatened but no data	[14] Not threatened	[15] Endemic to region	[16] Ref. no
Vertebrates	Global	IUCN	+	+	+	+		+						−	1–4
Primates Lemurs	Africa Madagascar & Comores	IUCN	+	+	+	+						+	+	−	5
Swallowtail butterflies	Global	IUCN	+	+	+	+						+		−	6
Dolphins, porpoises, & whales	Global	IUCN	+	+	+	+		+				+			7
Birds	Global	IUCN	+	+	N	+						+		−	8
Birds	Africa	IUCN	+	+	+	+		+		+	+			−	9
Amphibia & reptiles	Global	IUCN	+	+	+	+		+				+		−	10
Invertebrates	Global	IUCN	+	+	+	+	+					+		−	11
Waterfowl	Global	M–L			N		+			+				−	12
Cycads	Global	dIUCN			N	+	+			+				−	13
All	Thailand		+	+	+	+								−	14
All	Virginia, USA	ESA	+	+	+	+	+		+					+	15
All	Baltic	IUCN	+	+	+	+						+			16
Primates	Africa & Asia		+	+							+	+	+		17, 18
All	Americas	TNC	+	+	N	N							+		19
Plants	Australia	IUCN	+		+	N			+			+			20
Migrant birds	Neotropics	Rab			+	+	(+)								21
Insects	Britain	IUCN			+	N		+		+				+	22
Other invertebrates	Britain	IUCN			+	N		+		+		+		+	23

Column [3]: IUCN, IUCN criteria; M–L, Mace and Lande (1991); dIUCN, Mace et al. (1992); ESA, Endangered Species Act (Rohlf 1989); Rab., Rabinowitz et al. 1986, TNC, Master (1991)

Columns [4]–[15]: +, characteristic scored in threatened species categorization system; N, numerical guidelines given; −, not relevant

Column [16]: 1, IUCN (1972); 2, IUCN (1978); 3, IUCN (1979); 4, IUCN (1975); 5, IUCN (1985); 6, IUCN (1990); 7, IUCN (1985); 8, IUCN (1991); 9, Collar and Stuart (1985); 10, IUCN (1982); 11, IUCN (1983); 12, Green (1992); 13, Osborne (submitted); 14, TISTR (1991); 15, Terwilliger (1990); 16, Ingelof et al. (1993); 17, Oates (1986); 18, Eudeu (1987); 19, Master (1991); 20, Briggs and Leigh (1988); 21, Reed (1992); 22, Shirt (1987); 23, Bratton (1991) increase in the number of taxa (species or subspecies) listed as threatened (61 to 74 for cycads; 36 to 52 for waterfowl). However, this was not simply due to a shifting of boundaries, there was quite a substantial change in the set of taxa listed as threatened. In the cycads, 15 taxa were threatened according to IUCN definitions but not threatened by the quantitative criteria, and 20 were threatened according to quantitative criteria but not threatened by IUCN definitions. The equivalent figures for waterfowl are 27 and 11 taxa, respectively.

Almost all the systems also use small range size or number of sites explicitly, usually in a definition for a category such as 'rare'. Rarity can take a variety of forms and the framework suggested by Rabinowitz (1981) and Rabinowitz et al. (1986) is useful for considering extinction risk in relation to different forms of rarity (Thomas and Mallorie 1985; Rabinowitz et al. 1986). Species with small population sizes, restricted ranges, and narrow habitat specializations are always considered especially vulnerable, and the classically 'common' species with abundant populations over large geographical ranges and broad habitat types are always considered 'safe'. However, there is little consensus about the relative risks faced by taxa belonging to the other 6 forms of rarity (see Reed 1992; Kattan 1992). There also two practical problems which limit the use of these classifications at a general level. One is that each of these three variables is actually a continuum, and some arbitrary decisions have to be made about where the cut-off points lie. Secondly, there are always difficulties and compromises in the measurement of habitat specificity.

Several criteria relate in some way to management or its consequences and this is a difficult area in the categorization of threat. On the one hand, there are demands to have species listed as threatened, even if past or current management has led to a stable or increasing population. This is especially the case where threatened species categories are translated directly into legislation or into species protection. A logical consequence is that dependence upon protection would need to be a criterion for endangered status, and no species under management would ever leave the threatened species lists. On the other hand, an aim of conservation action should be to reduce the number of species listed as threatened. It would seem to be more logical to differentiate between threatened forms (i.e., those that are thought to be currently at high risk of extinction) and species dependent upon continuing management. This is not quite the same as the intent of the IUCN category 'out of danger' (now rarely used) which is used for cases where a species was previously listed but has recovered, and is therefore an historical rather than a continuing status assessment.

It seems surprising that population declines have not more commonly played a role in the listing of species, yet only 5 out of the 19 systems incorporate this in their definitions. Fundamentally, an endangered species in one that is showing or is expected to show evidence of decline. Combined with small population size or small range sizes, population decline seems intuitively to be a more reliable indicator of risk than do either of these two characteristics alone.

The categorization of species that have shown marked reduction in numbers or range from recent or historical levels is a reflection of a widespread concern that conservation should do more than prevent extinction. Functioning ecosystems should be maintained with their species complement at some level that may be well above that needed to prevent extinction. This view relies on the notion that there is some 'correct' or 'stable' level for each and every species, but it is very difficult to determine where this should be set. Historically and geologically, species numbers have always fluctuated and it is difficult to know how to implement this kind of approach practically. In any case, this would be an entirely different enterprise from the current one which is to limit the current extinction crisis.

In many published lists of threatened species a large number are classified as suspected of being threatened but with insufficient data to make a definite judgement. The category 'insufficiently known' in the IUCN definitions is specifically for this situation, and it has been widely applied. The classification is not useful for conservation planning since it is unclear where these species sit in relation to those that can be said definitively to have a high or moderate extinction risk, and it does not indicate anything about the kind of information that is required or that which is available. Conversely, taking species that are poorly known out of the threatened species lists altogether, may prejudice their survival simply on the basis that we know little about them. In addition, species may be classified as insufficiently known for a variety of reasons. They may be at risk because their only known habitat is being lost, but so little is known about their status that it is impossible to say whether they are 'endangered', 'vulnerable' or 'rare'. Alternatively, they may be known only from historical records from a remote site rarely visited by naturalists. Their actual status could be anything from extinct to common. Finally, Cooke (1991) gives a reasoned argument for a very broad application of the 'insufficiently known' category in the categorization of cetaceans. He argues that this category should be applied to species that are not definitely known *not* to satisfy the criteria for any of the other threatened categories, and that the only situation where this could be true would be where all or most populations are known *not* to be declining. This requires good information and among the cetaceans only one species (the grey whale) qualifies. Therefore, all others that are not classified in another threatened category are classified as insufficiently known. If this logic were applied to many groups the vast majority of species would be classified in this category.

Ideally, three things could help resolve the problem of how to treat poorly known species. First, the criteria for the categories could explicitly include risks derived from habitat change or loss affecting many species for which direct information on status is lacking. Second, classifications could be accompanied by some statement about the extent and reliability of information used to make the evaluation. Third, there could be a separate classification for species for which additional information is required before extinction risk can be evaluated. This would be distinct from the Red List and would more effectively highlight those species in need of survey or study, as opposed to those known to need protection. However, these measures alone do not solve the problem of how best to cope with the extent of our ignorance.

Few systems have explicitly included 'not threatened' in their categorization scheme, and as discussed earlier, this has led to uncertainty about the status of those species not listed. There seems to be no logical reason why this cannot be included, except for a reluctance be publish an assessment that could be so easily and disastrously proven wrong. If confidence can be placed in the criteria for threatened categories and a taxon can be shown to not qualify, then this should be stated. In fact, very often the information is so poor that the taxon cannot be shown *not* to qualify; which is a manifestation of the problem of how best deal with poorly known forms (see above). Undoubtedly, there should be a 'not threatened' or similarly named category, as there should be a category for taxa that have not even been evaluated against the definitions.

Finally, a number of regional studies have included categories that reflect the distribution of taxa inside and outside the region, especially to indicate those that are endemics, or whose major populations are found locally. This is generally more useful for conservation planning than assessing extinction risk (see below).

13.5   The proposed new IUCN criteria

The development of new IUCN categories is now well under way (Mace *et al.* 1992). These new definitions and criteria are still under active review and refinement, but an outline is presented here of version 2.1 (IUCN/SSC 1993).

The threatened species classification scheme falls within a larger scheme which will be applicable to all species and which will indicate whether a species has been assessed, whether there was sufficient information to categorize according to threat level and, if so, whether the species was felt to be threatened, not threatened or in need of continuing conservation management. The threat categories are defined only in terms of extinction risk, with decreasing threat levels over increasing time periods (Mace and Lande 1991). There are three threat categories ('critical', 'endangered', and 'vulnerable') which fall on a continuum, and a fourth category ('susceptible') which is reserved for taxa that do not qualify for any of the higher threat categories but which, as a consequence of restricted distributions are continually at risk of extinction. A series of quantitative criteria are provided for the categories of 'critical', 'endangered', and 'vulnerable', and to qualify for listing a species has to satisfy one of these. The five criteria are measures of: (1) observed, inferred or projected decline rates; (2) small populations that are either single or fragmented associated with an observed, inferred or projected decline; (3) small geographic range areas or extents associated with an observed, inferred or projected decline; (4) very small population sizes; and (5) a quantitative analysis predicting a given extinction risk within a specified time period (see Mace *et al.* 1992).

The system was designed to be appropriate for all macro fauna and flora, and although early drafts presented separate criteria for different major taxa, it became clear that with this approach species that had unusual life histories for their own major taxa might be judged by inappropriate criteria. It was therefore more conservative to consolidate criteria for diverse major lifestyles into a single set of criteria, and allow meeting any one to qualify the species for listing at that level. Depending on the perspective of the classifier, some of the criteria therefore may appear inappropriate or even absurd. However, under this system what matters is whether any of the criteria are met, not whether all are appropriate. The current review procedure has aimed to investigate whether application of the proposed criteria to diverse taxonomic groups indicates false listings and so far it has not appeared necessary to alter this structure, although there are concerns about some of the levels set.

13.6 Estimating extinction rates from threatened species lists

Because most lists are based on non-quantitative criteria and definitions, they cannot be used to make predictions about extinction rates. Smith *et al.* (1993*a*) have recently analysed the changes in species lists in IUCN Red Lists published between 1986 and 1990 to make some estimates of extinction risk. However, as they make clear, for most taxa these will be underestimates because of incomplete evaluations (see above). However, once the definitions for categories, and the criteria that determine listing under them, become quantitative, it will be possible to use these lists to make estimates of future extinction rates. A major caveat here is that listing of a taxon under a threat category does not necessarily constitute a prediction, because the very fact that it has been perceived to be in trouble, and placed on a Red List, should encourage effective conservation actions that reduce the extinction risk.

At this early stage in their development it is not appropriate to use the new draft IUCN criteria in this context. However, the quantitative definitions and criteria proposed by Mace and Lande (1991) have now been applied to a range of vertebrate taxa, mainly through activities of various IUCN Species Survival Commission Specialist Groups (Seal *et al.* 1993; Mace 1994). For a range of higher taxa, species and subspecies have been classified as 'critical' (50% risk of extinction in 5 years or 2 generations, whichever is longer), 'endangered' (20% risk of extinction in 20 years or 10 generations, whichever is longer), 'vulnerable' (10% risk of extinction in 100 years) or 'safe'. These assessments can be used to make some very rough estimates of future extinction rates, using a similar methodology to Smith *et al.* (1993*a*), except that here the analysis is based on evaluations of extinction risks across all extant members of certain higher taxa, instead of on recorded extinctions. The data presented in Table 13.2 are compilations from Mace (1994) and show the percentage of species in 10 vertebrate taxa placed in each of the threatened categories 'critical', 'endangered', and 'vulnerable'. Only species-level estimates are presented here, although in most taxa (not marsupials and canids) many classifications were made at subspecific level, and the species not then evaluated. The categories are defined by single risk and time points, but for the purpose of this analysis all were standardized to 100 years by fitting exponential extinction functions.

On this basis, no 'critical' taxa and only about 33% of 'endangered' taxa are expected to persist for 100 years. Applying these survival rates across threatened

Table 13.2 Extinction rates for vertebrate species calculated from threatened species
categories

	'Critical' (%)	'Endangered' (%)	'Vulnerable' (%)	No.	Est. % extinct in 100 yr	Est. yr to 50% extinction
Reptilia						
Boidae	5.9	12	35	17	17	365
Varandidae	0	3.5	34	29	6	1168
Iguanidae	4.0	8.0	56	25	15	428
Birds						
Anseriformes	4.6	8.3	20	109	16	404
Gruidae	17	0	50	6	19	335
Psittaciformes	7.3	8.3	24	302	15	421
Bucerotidae	10	30	40	10	34	166
Mammals						
Marsupiala	3.4	11	34	179	14	453
Canidae	5.9	12	21	34	16	403
Cervidae	29	29	21	14	50	101

Data on threat levels categorizations from workshop documents reproduced in CBSG
(1992)

classes gives the percentage of species in each group that are expected to be
extinct in 100 years time (Table 13.2). These percentages range from 6% to 50%.
The 50% figure for cervids may be rather inflated because most cervid taxa were
evaluated at the subspecific level, where the proportion of threatened forms was
lower (see Mace 1994). These values are similar to estimates of species extinction
rates derived from species–area curves. Recent estimates for tropical forest
species lie between 10% and 40% loss over 100 years (Reid 1992).

From these values, the characteristic extinction time, or estimated time to 50%
extinction, can be calculated, and these values are shown in the final column of
Table 13.2. They range from about 100 years to over 1000, but for most taxa are
around 300–400 years. These time periods are much shorter than those calculated
from recorded extinctions and at the very low end of estimates based on
transitions of species through existing Red List categories (Smith *et al.* 1993*a*).
Smith *et al.* (1993*a*) note that their estimates were likely to be low due to under-
recording, and the results of this analysis, where recording is complete, bear this
out.

There are, however, several reasons why these estimated times might be rather
too short. First, the definitions for these categories are based on time periods
measured in years or generations, and the generations measure will be used
whenever species generation lengths exceed 2.5 years (for 'critical') or 2 years (for
'endangered'). Most of the taxa in Table 13.2 have generation lengths
substantially longer than this. Second, there could be a bias from only using taxa
that were not evaluated at subspecific level, since these might commonly be

Table 13.3 Vertebrate extinction rates at different taxonomic levels and adjusted for generation length

Level	'Critical' (%)	'Endangered' (%)	'Vulnerable' (%)	No.	Est. % extinct in 100 yr		Est. yr to 50% extinction	
					Low	High	Low	High
Species	5.9	9.7	28	725	15	11	433	613
Subspecies	8.3	16	28	554	22	14	281	447

restricted range forms or island endemics, which would then be expected to have a higher extinction risk. In Table 13.3 the average values across all species and subspecies in this data set are analysed, and the effects of increasing generation length to 6 years (a rough estimate of a median) is shown. In fact, on average, the subspecies data give higher extinction risks (22% expected extinct in 100 years compared to 16% for species), and have a characteristic extinction time of 281 instead of 433 years. Increasing generation length increases the characteristic extinction time for species from 433 years to 613, and for subspecies from 281 to 447.

These estimates are still very crude. The criteria used to classify taxa into the categories are only very approximate (Mace and Lande 1991) and have not been, and probably cannot be, generally validated. Further, the procedure used to standardize them all to a 100-year period is simplistic. Also, the taxa were assessed in workshop sessions, and are inevitably based on very little information. There is undoubtedly a strong inclination to be highly conservative in making estimates under these conditions, especially as in most cases there was no option to place taxa into an 'insufficiently known' category, although some remained unclassified (Seal *et al.* 1993). They may therefore represent worst-case assessments. Finally, the results should not be generalized across other vertebrate taxa since the groups so far analysed probably represent higher order taxa that most clearly contain large numbers of threatened forms. No such analysis has yet been performed on, for example, rodents or passerine birds, and the outcome might be quite different.

The results are, however, interesting since they generally corroborate analyses made on species extinction rates from entirely independent methods: from species–area curves (Reid 1992) and from analyses of rates of movement of species through categories in existing Red Lists (Smith *et al.* 1993a).

13.7 The application of threatened species categories in setting priorities for conservation action

Interesting as the analysis of extinction rates is, the compilation of threatened species lists should not necessarily be an end in itself. In terms of conservation action aimed at limiting the extinction rate, the categories can be used in systems for setting conservation priorities, and determining appropriate short- and longer-term activities. In this section I review these schemes and the factors they incorporate.

Table 13.4 lists a number of priority classification schemes incorporating threat categories at particular taxonomic and/or regional levels. The schemes are quite diverse. Some have been developed by national or regional conservation authorities, others are global priorities for taxonomic groups prepared by Specialist Groups of the IUCN Species Survival Commission, and others are for very specific purposes. Equally diverse is the number of different characteristics that have been used in the setting of priorities. Not included in Table 13.4 are the algorithms that have been used to develop priority rankings, most of which translate scores on each of the characteristics into a single rank. In most cases, the threat category is dominant in the final rank. For example, in some cases (e.g., Oates 1986; Wager and Jackson 1993) there are simply more ranks for the threat classification than for other variables so that high-risk species will tend score highly overall. In others (e.g., Garnett 1992) the first ranking is by threat category and other variables are only brought in later on.

In a few of these examples, the threat category is the only character used, and is directly translated into a priority. More commonly, other features are scored. The most commonly applied variable after threat category is the association of the evaluated species with other species. All other things being equal it is appropriate to focus attention on taxa whose protection will benefit a greater number of other species, and although most of the systems used here are relatively simple, there is no reason why the more sophisticated algorithms developed recently (Pressey *et al.* 1993; Forey *et al.* in press) should not be used in this context. This procedure could also incorporate another commonly used characteristic, reflecting the taxonomic or genetic uniqueness of the taxon. Simplistically, this characteristic will favour species that are taxonomically distinct (e.g. a monotypic genus) but various kinds of taxonomic weighting schemes could be used here (Williams and Humphries in press).

A number of variables associated with high extinction risk are also commonly scored, such as small population size, restricted range or number of sites, declines, and specialized habits or habitats. Generally, these contribute to the threat level estimation. Protection status is a variable that should much more obviously be used in setting priorities than in evaluation of threat. At a planning level, it will be important to know to what extent different species at the same level of risk are already receiving protection of some sort, and in many classifications this is further subdivided into considerations of how effective this protection is (e.g., Briggs and Leigh 1988). The same is true of evaluations of the status of the species outside the region. In many cases, species threatened in one particular political region are secure elsewhere, and may simply be at the edge of their range. These species should rank lower than equally threatened forms that are endemic to the region, or that are globally threatened.

Recovery potential is an interesting variable, and again one that might be more

Table 13.4 Characteristics used in systems for evaluating conservation priorities

[1] Taxon	[2] Region	[3] Threat level	[4] Pop. size	[5] Range size/site number	[6] Pop. decline	[7] Protection status	[8] Recovery potential	[9] Association with other species	[10] Social/economic value	[11] Status out of region	[12] Ecological specialization	[13] Taxonomic or genetic value	[14] Poor data	[15] Ref. no.
Tortoises & freshwater turtles	Global	+												1
Old World fruit bats	Global	+				+								2
Canids	Global	+												3
Swallowtail butterflies	Global	+						+						4
Small carnivores	Global	+	+	+	+			+			+			5
Black rhino	Africa		+	+	+	+		+				+	+	6
Freshwater fish	Australia	+					+	+	+			+		7
Birds	Australia	+										+		8
Primates	Africa & Asia	+						+		+		+		9
All	Americas	+						+				+		10
Plants	Australia	+	+	+				+						11
Migratory birds	Neotropics		+	+							+			12
All	New Zealand	+	+	+	+	+			+		+			13
Birds	Americas	+	+	+		+							+	14
Vertebrates	Florida, USA	+	+	+	+		+			(+)	+		+	15

Columns [3]–[14]: +, characteristic scored in priority setting system

Column [15]: 1, Stubbs (1989); 2, Mickleburgh *et al.* (1992); 3, Ginsberg and Macdonald (1990); 4, New and Collins (1991); 5, Schreiber *et al.* (1989); 6, Cumming *et al.* (1990); 7, Wager and Jackson (1993); 8, Garnett (1992); 9, Oates (1986); 10, Master (1991); 11, Briggs and Leigh (1988); 12, Reed (1992); 13, Molloy and Davis (1992); 14, Collar *et al.* (1992); 15, Millsap *et al.* (1990)

frequently applied. However, its use poses some demanding analyses and raises some difficult questions to do with species triage. When should species with low recovery potential be given low priority on the basis that scant resources for conservation should be allocated to species that will respond, and should some 'hopeless cases' be written off altogether?

The analysis of species data for setting conservation priorities is an important developing area, and one that has so far received rather little attention. It is, however, one of the most significant applications of threatened species categories.

13.7   Conclusions

The development of more objective and systematic methods for evaluating the threat status of species will have many implications for conservation action. At one level, it will allow a better general evaluation of the extent of the current species extinction spasm, and what the regional and taxonomic biases are. At a practical level, however, it will permit the incorporation of threat levels into the rational planning of conservation actions.

Summary

Threatened species lists are widely consulted as sources of information on the status of species. However, their application to planning for conservation is limited because they have not been developed systematically, and because the criteria used to judge extinction risk are subjective. Recently, new proposals have been made which will increase the broader usefulness of these lists, and components of these proposals are reviewed here.

Apart from indicating the geographical and taxonomic groups containing most threatened species, these lists can provide other kinds of information. Some data from threatened species classifications using quantitative criteria provide a new method for estimating extinction rates in a variety of vertebrate taxa. This analysis suggests that over the next 100 years, the extinction rate could be as high as 15–20% in these groups. These values are comparable to those based up extrapolations from species–area curves.

However, allocating threatened species categories is only a first step towards developing rational systems for setting conservation priorities. These systems will need to consider a quite different set of variables, including those for incorporating species conservation priorities in area-based planning.

Acknowledgements

This work was supported by a grant from the Pew Scholars Program in Conservation and the Environment. The paper has arisen out of discussions with many people especially within the IUCN Species Survival Commission. In particular, I would like to acknowledge thoughtful input from Nigel Collar,

Justin Cooke, Kevin Gaston, Josh Ginsberg, Nigel Leader-Williams, Mike Maunder, E. J. Milner-Gulland, Simon Stuart, and Andrew Balmford.

References

Belovsky, G. E. (1987). Extinction models and mammalian persistence. In *Viable populations for conservation*, (ed. M. E. Souleé), pp. 35–58. Cambridge University Press.

Bratton, J. H. (ed.) (1991). *British Red Data Books, 3. Invertebrates other than insects.* Joint Nature Conservation Committee, Peterborough.

Briggs, J. D., and Leigh, J. H. (1988). *Rare or threatened Australian plants*. Australian National Parks and Wildlife Service, Canberra.

Burton, J. A. (1984). A bibliography of red data books. *Oryx*, **18**, 61–4.

CBSG. (Captive Breeding Specialist Group) (1992). *Conservation Assessment and Management Plan CAMP Summary Report*. IUCN/SSC/Captive Breeding Specialist Group, Minneapolis, Minnesota.

Collar, N. J. et al. (1992). *Threatened birds of the Americas. The ICBP/IUCN Red Data Book*. IUCN/ICBP, Cambridge.

Collar, N. J., and Stuart, S. N. (1985). *Threatened birds of Africa and related islands. The IUCN/ICBP Red Data Book*. IUCN/ICBP, Cambridge.

Cooke, J. (1991). Classification and reviews of species. In *Dolphins, porpoises and whales of the world. The IUCN Red Data Book*, (ed. M. Klinowska and J. Cooke), pp. 19–27. IUCN, Gland, Switzerland.

Cumming, D. H. M., de Toit, R. F., and Stuart, S. N. (1990). *African elephants and rhinos: status survey and conservation action plan*. IUCN, Gland, Switzerland.

Diamond, J. M. (1984). Normal extinctions of isolated populations. In *Extinctions*, (ed. M. Nitecki), pp. 191–246. University of Chicago Press.

Diamond, J. M. (1987). Extant unless proven extinct? Or extinct unless proven extant? *Conserv. Biol.*, **1**, 77–9.

Diamond, J. M. (1989). The present, past and future of human-caused extinctions. *Phil. Trans. Roy. Soc. London*, **B325**, 469–77.

Dietz, J. M., Dietz, L. A., and Nagagata, E. Y. (1994). The effective use of flagship species for conservation of biodiversity: the example of lion tamarins in Brazil. In *Creative conservation. Interactive management of wild and captive animals*, (ed. P. J. Olney, G. M. Mace, and A. T. C. Feistner), pp. 32–66. Chapman & Hall, London.

Eudey, A. A. (1987). *Action plan for Asian primate conservation*. IUCN, Gland, Switzerland.

Fitter, R., and Fitter, M. (ed.) (1987). *The road to extinction*. IUCN, Gland, Switzerland.

Forey, P., Humphries, C. J., and Vane-Wright, R. (1994). *Systematics and conservation evaluation*. Oxford University Press.

Garnett, S. (1992). *The action plan for Australian birds*. Australian National Parks and Wildlife Service, Canberra.

Ginsberg, J. R., and Macdonald, D. W. (1990). *Foxes, wolves, jackals and dogs. An action plan for the conservation of canids*. IUCN, Gland, Switzerland.

Green, A. (1992). Wildfowl at risk. *Wildfowl*, **43**, 160–84.

Imboden, C. (1987). From the Director's desk: Green Lists instead of Red Books? *World Birdwatch*, **9**, 2.

Ingelof, T., Andersson, R., and Tjernberg, M. (1993). *Red Data Book of the Baltic region*. Swedish Threatened Species Unit, Uppsala, Sweden.

IUCN (International Union for the Conservation of Nature and Natural Resources) (1972). *Red Data Book, Vol. I. Mammals*. IUCN, Cambridge.

IUCN (International Union for the Conservation of Nature and Natural Resources) (1977). *Red Data Book, Vol. IV. Fish*. IUCN, Cambridge.

IUCN (International Union for the Conservation of Nature and Natural Resources) (1978). *Red Data Book, Vol. II. Birds*. IUCN, Cambridge.

IUCN (International Union for the conservation of Nature and Natural Resources) (1979). *Red Data Book, Vol. III. Amphibia and reptiles*. IUCN, Cambridge.

IUCN (International Union for the Conservation of Nature and Natural Resources) (1982). *The IUCN Amphibia–Reptilia Red Data Book, Part I*. IUCN, Gland, Switzerland.

IUCN (International Union for the Conservation of Nature and Natural Resources) (1983). *The IUCN invertebrate Red Data Book*. IUCN, Gland, Switzerland.

IUCN (International Union for the Conservation of Nature and Natural Resources) (1985). *Threatened swallowtail butterflies of the world*. IUCN, Gland, Switzerland.

IUCN (International Union for the Conservation of Nature and Natural Resources) (1988). *Threatened primates of Africa. The IUCN Red Data Book*. IUCN, Gland, Switzerland.

IUCN (International Union for the Conservation of Nature and Natural Resources) (1990). *1990 IUCN Red List of threatened animals*. IUCN, Gland, Switzerland.

IUCN (International Union for the Conservation of Nature and Natural Resources) (1991). *Dolphins porpoises and whales of the world. The IUCN Red Data Book*. IUCN, Gland, Switzerland.

IUCN/SSC (1993). *Draft IUCN Red List Categories, version 2.1*. IUCN, Gland, Switzerland.

Kattan, G. H. (1992). Rarity and vulnerability: the birds of the Cordillera Central of Colombia. *Conserv. Biol.*, **6**, 64–70.

Losos, E. (1993). The future of the US Endangered Species Act. *TREE*, **8**, 332–6.

Mace, G. M. (1994). An investigation into methods for categorising the conservation status of species. In *Large scale ecology and conservation biology*, (ed. P. J. Edwards, R. M. May, and N. R. Webb), pp. 295–314. Blackwell, Oxford.

Mace, G. M., and Lande, R. (1991). Assessing extinction threats: toward a reevaluation of IUCN threatened species categories. *Conserv. Biol.*, **5**, 148–57.

Mace, G. M., *et al.* (1992). The development of new criteria for listing species on the IUCN Red List. *Species*, **19**, 16–22.

Master, L. (1991). Assessing threat and setting priorities for conservation. *Conserv. Biol.*, **5**, 559–63.

May, R. M. (1988). How many species are there on earth? *Science*, **241**, 1441–9.

Mickleburgh, S., Hutson, A. M., and Racey, P. A. (1992). *Old World fruit bats: an action plan for their conservation*. IUCN, Gland, Switzerland.

Millsap, B. A., Gore, J. A., Runde, D. A., and Cerulean, S. I. (1990). Setting priorities for the conservation of fish and wildlife species in Florida. *Wildlife Monogr.*, **111**, 1–57.

Molloy, J., and Davis, A. (1992). *Setting priorities for the conservation of New Zealand's threatened plants and animals*. Department of Conservation, Wellington, New Zealand.

Munton, P. (1987). Concepts of threat to the survival of species used in Red Data Books and similar compilations. In *The road to extinction*, (ed. R. Fitter and M. Fitter), pp. 72–111. IUCN, Gland, Switzerland.

New, T. R., and Collins, N. M. (1991). *Swallowtail butterflies: an action plan for their conservation*. IUCN, Gland, Switzerland.

Oates, J. F. (1986). *Action plan for African primate conservation: 1986–1990*. IUCN, Gland, Switzerland.

Osborne, R. (1994). The 1991–1992 world cycad census and a proposed revision of the

IUCN threatened species status for cycads. In *Proceedings of the third international conference on cycad biology* (ed. P. Vorster), in press.

Pimm, S. L. (1992). *The Balance of nature*. University of Chicago Press.

Pressey, R. L., Humphries, C. J., Margules, C. R., Vane-Wright, R. I., and Williams, P. H. (1993). Beyond opportunism: key principles for systematic reserve selection. *TREE*, **8**, 124–8.

Rabinowitz, D. (1981). Seven forms of rarity. In *Biological aspects of rare plant conservation*, (ed. H. Synge), pp. 205–17. Wiley, Chichester.

Rabinowitz, D., Cairns, S., and Dillon, T. (1986). Seven forms of rarity and their frequency in the flora of the British Isles. In *Conservation biology: the science of scarcity and diversity*, (ed. M. E. Soulé), pp. 182–204. Sinauer, Sunderland, MA.

Reed, J. M. (1992). A system for ranking conservation priorities for neotropical migrant birds based on relative susceptibility to extinction. In *Ecology and conservation of neotropical migrant landbirds*, (ed. J. M. Hagan, III and D. W. Johnston), pp. 524–36. Smithsonian Institution Press, Washington, DC.

Reid, W. V. (1992). How many species will there be? In *Tropical deforestation and species extinction*, (ed. T. C. Whitmore and J. A. Sayer), pp. 55–73. Chapman & Hall, London.

Rohlf, D. R. (1989). *The Endangered Species Act: a guide to its protections and implementation*. Stanford Environmental Law Society, Stanford, California.

Schreiber, A., Wirth, R., Riffel, M., and Rompaey, H. V. (1989). *Weasels, civets, mongooses and their relatives. An action plan for the conservation of mustelids and viverrids*. IUCN, Gland, Switzerland.

Scott, J. M., Csuti, B., Jacobi, D., and Estes, J. (1987a). Species richness: a geographical approach to protecting future biodiversity. *Bioscience*, **37**, 782–8.

Scott, P., Burton, J. A., and Fitter, R. (1987b). Red Data Books: the historical background. In *The road to extinction*, (ed. R. Fitter and M. Fitter), pp. 1–5. IUCN, Gland, Switzerland.

Seal, U. S., Foose, T. J., and Ellis-Joseph, S. (1993). Conservation assessment and management plans (CAMPs) and global captive action plans (GCAPs). In *Creative conservation—the interactive management of wild and captive animals*, (ed. P. J. Olney, G. M. Mace, and A. T. C. Feistner). Chapman & Hall, London.

Shirt, D. B. (ed.) (1987). *British Red Data Books*, 2. *Insects*. Nature Conservancy Council, Peterborough.

Smith, F. D. M. *et al.* (1993a). Estimating extinction rates. *Nature*, **364**, 494–6.

Smith, F. D. M. *et al.* (1993b). How much do we know about the current extinction rate? *TREE*, **8**, 375–8.

Stubbs, D. (1989). *Tortoises and freshwater turtles. An action plan for their conservation*. IUCN, Gland, Switzerland.

Talbot, L. M. (1959). *A look at threatened species*. IUCN, Gland, Switzerland.

Terwilliger, K. (1990). *Virginia's endangered species: proceedings of a symposium*. Department of Game and Inland Fisheries, Virginia, USA.

Thomas, C. D., and Mallorie, H. C. (1985). Rarity, species richness and conservation: butterflies of the Atlas Mountains in Morocco. *Biol. Conserv.*, **33**, 95–117.

TISTR (Thailand Institute of Scientific Research) (1991). *Endangered species and habitats of Thailand*. TISTR, Thailand.

Wager, R., and Jackson, P. (1993). *The action plan for Australian freshwater fishes*. Australian Nature Conservation Agency, Canberra.

WCMC (World Conservation Monitoring Centre) (1992). *Global Diversity: Status of the Earth's Living Resources*. Chapman & Hall, London.

Williams, P., and Humphries, C. J. (1994). Biodiversity, taxonomic relatedness and endemism in conservation. In *Systematics and conservation evaluation*, (ed. P. Forey, C. J. Humphries, and R. Vane-Wright), pp. 15–33. Oxford University Press.

14

The scale of the human enterprise and biodiversity loss

Paul R. Ehrlich

14.1 Introduction

Since every species and population can be assigned intrinsic (and perhaps practical) value, we should be concerned with both the number and rate of extinctions (P. R. Ehrlich and A. H. Ehrlich 1981, 1992). It also seems reasonable to place value on the proportion of a higher taxon that is disappearing: the annual loss of 1000 species from the class Aves would be considered as substantially more serious than the loss of 1000 species from the class Insecta, and the extinction of the only remaining coelocanth would be deemed a scientific tragedy.

Here, however, I will limit my focus to population and species extinctions, and start with a plea to give adequate attention to the former (Ehrlich and Daily 1993; Mace, Chapter 13). It is easy for the naïve to imagine there is no species extinction crisis, and complex analyses are required to show that the situation could actually be extremely serious (e.g., Smith *et al.* 1993*b*). But little sophistication is required to see that populations are disappearing at a high rate, and that will have disastrous consequences (Ehrlich and Daily 1993) regardless of the fate of species. After all, if population extinctions reduced all remaining species to single minimum viable populations, no further species extinctions would have occurred. Nonetheless, an extinction catastrophe would have taken place that might well, through interruption of ecosystem services (Holdren and Ehrlich 1974; P. R. Ehrlich and A. H. Ehrlich 1981), cause the demise of humanity as well.

Directly obtaining rates of loss of species and of populations requires overcoming a daunting array of problems. First, no one is sure how much diversity there is, however defined (e.g., May 1988). At the species level, estimates range from 5 million to 30 million or even 100 million (e.g., Wilson 1992), and there probably are billions of distinct populations (Ehrlich and Daily 1993; Daily and Ehrlich, in prep.). Needless to say, even knowing present absolute extinction rates would be of little use in obtaining proportional rates of extinction when there is roughly an order of magnitude uncertainty on the size of the stock.

Biologists should initiate steps to narrow the uncertainties about the magnitude of biodiversity and to compute present loss rates for a stratified sample of taxa. These tasks are so enormous, however, that it behoves us to examine indirect ways of determining the rate of decay of biodiversity. After all, it is not necessary to have counted, named, and established measures of similarity among the grains of sand, pebbles, shells, and rocks on a beach in order to determine for practical purposes how rapidly the beach is eroding.

Here, the following propositions are evaluated to see whether an indirect index of the rate of biodiversity loss can serve in place of direct estimates that are not available now and are unlikely to be available in the critical decades ahead:

1. Rates of extinction of both populations and species are related to the rate of habitat loss.

2. The rate of habitat loss increases with the scale of the human enterprise.

3. Total energy use is a reasonable, if imperfect, surrogate for the scale of the human enterprise and its environmental impact. Thus, that use is positively correlated with extinction rate, and a doubling of energy use leads to roughly a doubling of the rate. By applying data on increase in energy use to historical data on extinction rates, current rates can be estimated.

14.2   Proposition 1: extinction tracks habitat loss

No organisms occur in all habitats, and most have quite narrow habitat requirements. Plants often require characteristic soils, regimes of temperature, humidity, and light. In one sample of British plant species, about half were restricted to one or a few specialized habitats (Rabinowitz et al. 1986). Members of the most speciose taxonomic group, phytophagous insects, each tend to feed on relatively few plants (e.g., Futuyma 1991; Bernays and Graham 1988; Ehrlich and Murphy 1988).

So, when habitats are destroyed, populations and, eventually, species inevitably go extinct. When the sand dune habitat of the butterfly *Cercyonis sthenele sthenele* was developed as San Francisco grew in the last century, that distinctive population (subspecies) disappeared. When the mature swamp forest habitats of the south-eastern United States were fragmented, the ivory-billed woodpecker which required large tracts of such habitat, lost population after population and finally went extinct. In Australia I was once told by a forester not to worry about the destruction of Queensland's tropical forests 'because they'll grow back.' That is a highly dubious proposition for the plants unless sources of propagules are preserved and appropriate microclimates for their germination and growth (often provided by the forests themselves) can be restored. But it is rank nonsense for organisms like parrots and canopy butterflies, which cannot remain in orbit for the many decades needed to re-establish their habitat.

14.3   Proposition 2: habitat loss increases with the human enterprise

Determining the overall rate of habitat loss with precision is very difficult for several reasons. First, comprehensive numbers on even total destruction of habitats such as clear-cutting of forests, plowing of grasslands, and draining of swamps, are not available. Existing statistics are limited to some habitats in some nations, leaving little basis for making 'bottom up' global assessments (e.g., WRI 1990, 1992).

The figures that are available do give a feel for how extensive such damage can be. Percentages of total forest area that has been lost are available for a sample of 40 African nations and range from 30% in Zambia to 91% in The Gambia, with an average of 68%. Losses in 14 sampled Asian nations range from 34% to 96%, with an average of 69%. China, largely deforested, is not included; India has lost some 78% of its forests. Unhappily, estimates are only available for four western hemisphere nations: United States, 26%; Argentina, 50%; Guatemala, 60%; and Mexico, 66% (WRI 1990).

The extent of total habitat destruction varies among habitat types and regions. Nearly all the old-growth forests of the Pacific Northwest of the United States are gone. Approximately half of all tropical moist forests have been cut since 1950. Other kinds of habitat have suffered similarly. In the United States, virtually 100% of natural grasslands have been lost since 1492. Both Germany and the contiguous 48 United States have lost half their wetlands. About three-quarters of the coastal mangrove wetlands of India, Pakistan, and Thailand are gone (WRI 1992; Brown et al. 1992).

Second, much habitat that is not destroyed is 'lost' to some organisms through various forms of habitat conversion short of outright destruction. For example, converting many old-growth pine forests in the southeastern United States to even-aged stands of pines made them uninhabitable by red-cockaded woodpeckers, but had less impact on various species of warblers.

Third, orbital or airborne sensors cannot provide comprehensive estimates of rates of habitat destruction. Remote sensing might seem an ideal solution to the precise measurement of those rates, and it can be useful in measuring deforestation, desertification, and other dramatic forms of habitat destruction. But remote sensors have a variety of limitations. One is that they often cannot make subtle but important discriminations. For instance, over forests they tend to measure surface (crown) characteristics, and thus they may not be able to detect major changes in forest biomass that do not disrupt the canopy significantly. Synthetic aperture radar is incapable of distinguishing different amounts of biomass above a level that excludes from analysis over half of Earth's vegetated area, including all mature tropical moist forests (Imhoff 1993).

Remote sensing is now of little value in detecting critical alterations of freshwater habitats. One study of New Zealand stream habitats found the fish communities in most sites depauperate in both species diversity and abundance (Swales and West 1991), a condition not apparent to many indirect means of assessment. The habitat changes responsible were clearance of native forests, drainage of wetlands, and river channel works for hydropower, agricultural development, and other uses. In theory, data acquired by remote sensing might, given the proper algorithms, eventually be used to determine extinction rates in lakes, rivers, and streams caused by such changes—if the remote sensing could distinguish native from exotic forest (since streams in the latter had few fish species).

On the other hand, habitat changes caused by introductions of exotics may never be detectable remotely, and invasions account for some 20–40% of known species extinctions (WRI 1989). Consider the dramatic modification of the Lake Victoria habitat that occurred with the introduction of Nile perch (*Lates niloticus*) which promptly devoured most of the endemic cichlid species (Lowe-McConnell 1987), or the introduced brown snake that virtually wiped out the avifauna of Guam. Several individuals of the snake have now been found on the runways at Honolulu International Airport (apparently having dropped from aircraft wheel wells). They pose a severe threat to the remnants of the distinctive endemic avifauna of the Hawaiian Islands, but would not show up in satellite images. Indeed, invasions are one of the most insidious and potent forms of habitat conversion, and one of the most difficult to evaluate from afar (Drake *et al*. 1989).

Habitat can be 'lost' in other very subtle ways, difficult to detect even on-site. For instance, even minor incursions into forest that do not significantly change vegetative structure can be catastrophic for larger forms of wildlife. As Terborgh (1983) put it: 'A skilled hunter equipped with a standard one-shot 16 gauge shotgun can single-handedly eliminate large birds and mammals within a radius of several hours' walk from his dwelling'. Before Terborgh found the undisturbed Cocha Cashu in the Peruvian Amazon, he, like most of us, did not realize that a tropical forest could have abundant mammals (especially primates).

In Costa Rica, hunting pressure has removed agoutis (*Dasyprocta punctata*) and crested guans (*Penelope purpurascens*) from many remnant forest patches. But extinction patterns are often complex—in southern Costa Rica, despite heavy hunting pressure, guans persisted in a recently converted (18–30 years ago) agricultural landscape with small (*c*. 3–30 hectares) forest patches from which even many forest butterfly species had already been eliminated (Daily and Ehrlich, in press).

In addition to these difficulties, data acquisition and analysis on the important conditions and processes to which remote sensing *is* ideally suited (such as rates of change in land-use patterns) fall far behind need (e.g., Myers 1989).

Indicators of conversion

Lacking global (or even national) assessments of most habitat destruction, scientists must develop and validate indirect measures of habitat conversion rates. One of the better indicators of such conversion seems to be change in the distribution of butterflies. In a few areas, because of the interest of amateur collectors, reasonably good historical records are available. Butterflies are probably preferable to birds because of the tight ties of butterflies to plants and their frequent occurrence as localized populations.

The best-known butterfly fauna world-wide is that of Britain, and its history is well summarized elsewhere in this volume (Thomas and Morris, Chapter 8; see also Heath *et al*. 1984). Of 59 species resident 150 years ago, 4 (7%) have gone extinct, and another 18 (30%) have undergone major range contractions (an average of some 40% loss of area). Climatic shifts are involved in some observed contractions and expansions of range, but recent changes in landscape management are clearly the major reasons for the declines (see Thomas and Morris, Chapter 8).

Studies of butterflies and other insects suggest that human impacts on populations of invertebrates (and probably small vertebrates) can push extinction rates higher than would be indicated by standard species–area relationships. Many species appear to have a metapopulation structure in which stochastic extinction is common and recolonization is necessary to allow the species to persist (Ehrlich and Murphy 1987; Svensson 1985). Reduction of the number of colonies can diminish the probabilities of recolonization of 'empty' habitat patches to the point where regional or global extinction follows (Lipscomb and Jackson 1964). This can occur long before total habitat area is reduced to the 10% that species–area relationships suggest is required to produce a 50% chance of extinction.

Population extinctions

Trying to determine rates of population extinctions is even more difficult than establishing rates of species extinctions. Populations are more difficult to define than species, and to my knowledge no systematic attempt is made anywhere to keep track of population extinctions in any major taxonomic group.

There is one data set that casts some light on the extinction of butterfly populations (and thus, perhaps, on the extinctions of herbivorous insects in general). The British butterfly distributions mapped by Heath *et al.* (1984) are based on records of presence or absence in a national grid consisting of 3600 10 km squares covering the entire British Isles. The maps show records of pre-1940 presence gathered from the literature (and showing, in combination with other records, the maximum historical extent of each species), records from 1940–69 gathered from field notes and collections by early contributors in a comprehensive scheme to determine which grid squares had which species, and field records as they were received by those collaborating in the intensive survey between 1970 and 1982. Thanks to the enthusiasm of British lepidopterists, while the maps are admittedly incomplete, 98% of the grid squares had been surveyed to some extent.

Using these maps, the shrinkage in number of populations in 54 of the 62 British butterfly species can be estimated. I excluded 3 migratory species, 2 widespread agricultural pests, and 3 species that had contracted and then re-expanded their ranges to equal or greater area. The exclusions did not change the picture significantly, but did reduce the average maximum number of squares occupied from 927 to 807. The results are given in Table 14.1.

One assumption made in Table 14.1 is that all squares were occupied in prior periods—if there were butterflies in a square in 1978, butterflies were assumed to have been there in 1938. This is certainly not exactly true, but since the three species that were noted as expanding their ranges were excluded, it seems a

Table 14.1 Decline of British butterfly populations as indicated by records from grid squares

	Total all periods	Extinct by 1940	Extinct by 1969	Extant in 1970–82
Av. no. squares occupied	807	68	107	632
Av. percentage per species	100%	23%	17%	60%

reasonable assumption for the level of conclusions drawn here. A more interesting issue is how square occupancy corresponds to population number (Daily and Ehrlich, in prep.). Here, it is simply assumed that if a species disappeared from 50% of the grid squares, it lost 50% of its populations. That is almost certainly a conservative assumption, since most British species have geographically restricted, closed populations. Consequently, some species are known to be still present in grid squares although one or more populations have gone extinct in those squares. For instance, five populations of *Hesperia comma* have disappeared from squares in Dorset that still contain extant colonies and thus remain 'occupied' on the grid square maps, and the same is true of virtually all other Dorset butterflies (Thomas and Webb 1984).

Based on those assumptions, the loss of population diversity of British butterflies is clearly much more extensive than that of species diversity, as Table 14.1 shows.

14.4 Proposition 3: total energy use is correlated with extinctions

One can view the impact (I) of humanity on Earth's life support systems as the product of three factors, the size of the human population (P), per capita affluence (A) (as measured by per capita consumption), and an index of the environmental damage of the technologies (T) used to supply each unit of consumption (Ehrlich and Holdren 1971; Holdren and Ehrlich 1974; Ehrlich *et al.* 1977; P. R. Ehrlich and A. H. Ehrlich 1990). The $I = P \times A \times T$ identity is important as a heuristic device, but since only P is reported in national statistics, the identity's usefulness in analysis depends on estimating A and T. That problem can be largely solved by substituting per capita energy use (for which numbers are generally available) for $A \times T$.

Energy use is involved in virtually all of humanity's most environmentally destructive activities. Energy powers the automobiles, airplanes, electric lights, heaters, airconditioners, and household appliances of the rich, and is heavily involved in their manufacture, as well as in the construction of homes, office buildings, freeways, dams, and shopping malls. Mining and petroleum drilling are energy-intensive, as are the production of plastics, pesticides, and fertilizers, and production and distribution in intensive (industrial) agricultural systems. Over-harvesting of fuelwood and other biomass, which supplies an estimated 35% of energy consumed in poor nations, is a primary cause of deforestation and desertification. Chainsaws and bulldozers use energy, as does the modern transportation that makes it practical for rich nations to take easy advantage of the natural resources of the entire globe and to transport invasive organisms from continent to continent.

Brown snakes are not the only destructive organisms moved by energy-intensive vehicles. Hunters can now travel in jet aircraft and light planes (the latter much used by polar bear and wolf hunters and eagle poachers in North America), off-road vehicles, and power boats. Not all of their activities are destructive of biodiversity (in the United States, for instance, duck hunters may be, overall, a force for conservation), but many are. Bird-egg thieves in Britain make extensive use of mechanized transport; so do plant poachers. One of the last two populations of an endangered cactus was flown into Germany in 1978 in 15 suitcases, and truck-borne cactus rustlers harvest millions of cacti a year in the southwestern United States, making them rare near roads (P. R. Ehrlich and A. H. Ehrlich 1981). Intercontinental jet air traffic also makes the illegal trade in parrots, primates, and other wildlife much easier than it would be if only ships were available for transport. Therefore, although not perfect, in part because of regional differences in energy efficiency, per capita energy use seems a reasonable surrogate for per capita impact (P. R. Ehrlich and A. H. Ehrlich 1991).

Even where habitat loss can be quantified, determining its relationship to energy use can be very difficult. Several studies suggest that local or regional population growth is not a major trigger of rapid tropical deforestation (e.g., Kummer 1991). But external factors, such as demand for timber and beef in rich nations and the industrialization of agriculture (often to grow produce for sale to the wealthy), leading to the invasion of tropical forests by displaced farm labourers, clearly are related to the global scale of energy use.

One can hope, then, that changes in global energy consumption might average over the manifold regional differences and provide a satisfactory general index of extinction rates. To test this idea, consider the increase of extinction rates over background levels that would be predicted from the expansion of the human enterprise as measured by energy consumption. An estimate of the average species 'lifespan' of higher vertebrates is 200 000 to 2 million years (Ehrlich et al. 1977), giving a background extinction rate of species between 0.2 and 2 species per million species per year. This estimate brackets that of Raup (1988) and Wilson (1992) of 1 species per million species per year, which will be used here as the best available.

I assume that the impact of the human enterprise did not begin to move the extinction rate out of the background range before the agricultural revolution roughly 10 000 years ago. An exception to this might be found among the large mammals, because of the widespread extinctions of the Pleistocene megafauna in which *Homo sapiens* is almost certainly implicated (Martin and Wright 1967).

Megavertebrates, however, are a tiny portion of biodiversity, and the invention of agriculture was clearly the step that put our species on the road to being a global force.

Just before the agricultural revolution, there were perhaps 5–10 million people (Biraben 1979). A conservatively high guess of their per capita energy consumption would be about 0.2 kilowatts (kW), consisting mostly of wood-burning. The total impact of humanity then was, therefore, on the order of 0.001–0.002 terawatts (1 TW $= 10^{12}$ watts). Under those assumptions, human impact had multiplied roughly 7000- to 13 000-fold by 1990, when total energy consumption reached about 13 TW. Wilson's (1992) rather conservative appraisal based on island biogeographic theory is that human activity has increased the species extinction rate in rainforests (by reduction in area) between 1000- and 10 000-fold. So, those estimates overlap, and a reasonable (but not conservative) impact-based evaluation would be the loss today of about 10 000 species per million species per year.

Since 1600, a very conservative estimate of the number of bird and mammal species that have gone extinct is 175 (Smith et al. 1993a). That is an annual rate of roughly 30 species per million species in those taxa. From 1986 to 1990 15 bird and mammal species disappeared, a rate of over 250 per million species per year. That rate is about an order of magnitude lower than Wilson's (1992) conservative appraisal that rainforest species (of all groups) are disappearing at a yearly rate of 2700 per million species, and even further below the 10 000 per million (1%) per year implied by the above calculation on human impact. Thus, from the viewpoint of species extinction, the increased scale of the human enterprise suggests rather higher overall rates than the admittedly conservative estimates based on known vertebrate extinction rates. The correlation of species extinction with that scale would be stronger if insect species were currently suffering a relatively high rate of extinction. That would be occurring if in the extremely rich, poorly known, and unmonitored fauna of tropical moist forests many insect species had relatively restricted distributions.

If we use the rate of increase of total energy use from 1970 to 1990 (2.3% annually; Holdren 1991) as an index to future rates of extinction (-2.3% of species or populations annually), then the time to extinction of half of all species or populations is about 30 years. This figure, which is based on a constant exponential decay of diversity, seems quite high for species compared to other evaluations, but it could be low for populations.

In Britain, is there a relationship between energy consumption and the extinction of species and populations? Between 1925 and 1955, commercial energy consumption increased roughly 40%, but between 1955 and 1975 it went up only about another 5%. Over the entire 1925–82 period, it climbed about 50% (Darmstadter 1971; UN 1988). If population extinction rates were proportional to regional energy use, one would expect a loss of about 33% of populations in the 1925–82 period.

Assuming these UK energy figures are adequate surrogates for the scale of the human enterprise in the British Isles, since 1955 the rate of butterfly *population* extinctions in that nation has been faster than that of increase in human activities, while that of *species* extinctions has lagged (since only 1 species, 2% of the fauna, has disappeared in the last 40 years). The observation that extinction rates for populations are much higher than species extinction rates (as one would assume on general principles) certainly conforms with massive anecdotal data from continental Europe (e.g., Kudrna 1986) and other parts of the temperate zones. Extinctions of butterfly populations in North America, north of Mexico, also have been observed on numerous occasions. Most populations recorded by early collectors in the Los Angeles Basin are long gone, and my research group has observed the extinction of several populations of the Bay checkerspot butterfly, *Euphdydryas editha bayensis* over the roughly three decades that we studied it (Ehrlich *et al.* 1980). In contrast, no butterfly *species* has gone extinct in North America in historic times, although two distinctive subspecies in the San Francisco area were extirpated by urban expansion, one around 1880 and the other in the early 1940s (Scott 1986).

Similarly, many butterflies are declining in Japan, and population extinctions are common. But none of the 238 resident butterfly species has yet wholly disappeared from the Japanese islands (Sibitani 1989). In fact, to my knowledge there has been no well-documented extinction of a continental butterfly species anywhere in the world. The complete disappearance of the Italian endemic *Polyommatus exuberans*, would be an example if it actually was a distinct species (Kudrna 1986). It is likely that some species have disappeared unheralded in tropical areas (such as the unique *Styx infernalis* of the Chanchamayo Valley in the Peruvian Andes).

14.5 Interpretation

Obviously, attempting to compare rates of extinction with rates of habitat destruction involves dealing with incomplete information and pyramiding assumptions. On one hand, data on species extinctions in even the most thoroughly monitored groups are at best incomplete (e.g., Diamond 1987), and pathetically little information of any sort exists on population extinctions. On the other hand, overall evaluations of habitat destruction involve extrapolation from data on energy use as a surrogate for the scale of the human enterprise, then assume that scale to be well correlated with the loss of habitat, and finally assume that loss of habitat is closely related to rates of extinction. With the exception of the relationship between habitat loss and population extinction, all these assumptions are rather over-simplified.

Nonetheless, estimates based on them seem to fall between those generated by other methods on species and population extinctions. The power function relationship between habitat destruction and species extinction on one hand, and the cascading effects of disrupting metapopulations on the other, could explain this. We know on both theoretical and observational grounds that reducing a habitat area by half does not lead to the extinction of half the species in it. We also know that the loss of key habitat patches in a metapopulation system can cause the disappearance of all component populations.

14.6   Conclusion

If the human enterprise expands to anything roughly like the scale envisioned by the Brundtland report (WCED 1987), it will increase at least tenfold, to a 130 TW (1 TW $= 10^{12}$ watts) world, before growth stops. As a thought experiment, a 130 TW world today could imply, based on numbers in Smith *et al.* (1993*a*, *b*) for the actual current 13 TW world, only about 30 years to the extinction of half of all species of birds and mammals, and about 7 years for half of all species of palms. On the other hand, if the world were instantaneously transformed into the 30 TW world of the optimistic Holdren scenario (Holdren 1991), then those times to 50% extinction would be extended to about 130 years and about 30 years. Compare these with the global estimate of 30 years to 50% extinction based on current growth rates of energy consumption.

These projections are not reassuring. Furthermore, ecological (including climatic) systems often display non-linearities, long time-lags, and threshold effects that could make such projections conservative. For example, many tropical pastures contain isolated rainforest trees that were left standing to shade cattle and now are utilized by forest birds. They have limited lifespans and are not being replaced as they die. When they go, they may trigger a surge in population (and perhaps species) extinctions. Moreover, neither species nor population extinctions occur independently (e.g., Paine 1966; Daily *et al.* 1993). Consequently, extinction cascades may become commonplace in the next century.

Furthermore, *Homo sapiens* now threatens to alter the climate dramatically and already uses, co-opts, or destroys roughly 40% of terrestrial net primary productivity, the food supply of virtually all animals and most microorganisms that live on land (Vitousek *et al.* 1986). Finally, there is no guarantee that technological changes (as have already occurred in forest harvesting techniques) will not accelerate the expansion of the human enterprise and accelerate the exponential rate of diversity decay. In the light of such considerations, it seems only sensible to conclude that projections based on past and current rates of anthropogenic extinction could easily be low. In my view, there is ample justification for society to pay substantial premiums to insure against the loss of a large fraction of biodiversity.

14.7   Will refining knowledge of extinction rates make much difference?

Not *scientifically*. Rates are now so far above background that a major extinction episode is clearly underway. Biologists and others who appreciate the values of biodiversity need no more incentive to take action and to urge action. It is not even clear that substantial efforts to refine knowledge of overall species and population extinction rates are *scientifically* justified. Putting effort into that task might, for example, decrease that available for improving understanding of the scope and dynamics of biodiversity before most of it is gone (Ehrlich 1964; May 1988), for developing good 'rapid evaluation' methods for determining the geographic distribution of biodiversity so resources for conservation can be better allocated (Daily and Ehrlich 1994), and for focusing attention on preserving ecosystems, where effort obviously is needed.

On the other hand, better understanding of biodiversity losses would be useful *politically*. Carefully documented rates of population and species extinction in a sample of groups could help to counter lunatic claims such as that extinction rates are 'less than one-thousandth as great as doomsayers claim' (Simon and Wildavsky 1993). Better understanding of both the distribution and dynamics of the more spectacular elements of biodiversity can also make the tasks of conservation biologists easier. Public opinion on issues related to extinctions is unlikely to be swayed as much by scientific study as by publicizing the declines of charismatic megavertebrates—declines which can be employed to win legal protection for entire ecosystems. Lay people can relate well to pandas and whales; island biogeography of insects seems beyond the grasp of even some professors of business and political science.

Summary

This chapter evaluates the proposition that rates of population and species extinction can be assessed using an indirect measure: total consumption of energy (industrial plus traditional) by humanity. This proposition rests on three assumptions. First, the rate of extinction is proportional to the rate of habitat destruction because most organisms are adapted to rather limited environments. Second, the rate of habitat destruction is correlated with the scale of the human enterprise—the product of the number of people, per capita consumption, and the environmental damage done by the technologies used to supply each unit of consumption. Third, per capita energy use can be used as a surrogate for the latter two factors, consumption × technology. Total energy use is therefore an indicator of trends in extinction rates, and thus could be used to estimate the rates themselves. I examine these premises and conclude that they are sufficiently well-supported for biologists to use total energy consumption as an index of global extinction rates. That index however, is not useful politically because the assumptions upon which it is based are not understood by decision-makers and the general public.

Acknowledgements

I thank Gretchen Daily, Anne Ehrlich, and Fraser Smith (Center for Conservation Biology, Stanford), and Robert May (Department of Zoology, Oxford) for helpful comments on the manuscript. This work was supported in part by a grant from the W. Alton Jones Foundation.

References

Bernays, E., and Graham, M. (1988). On the evolution of host specificity in phytophagous arthropods. *Ecology*, **69**, 886–92.

Biraben, J.-N. (1979). Essai sur l'eévolution du nombre des hommes. *Population*, **34**, 13–25.

Brown, L. R. *et al.* (1992). *State of the world*. W. W. Norton & Co., New York.

Darmstadter, J. (1968). *Energy in the world economy*. Johns Hopkins, Baltimore.

Daily, G., and Ehrlich, P. Preservation of biodiversity in small rainforest patches: a rapid evaluation using butterfly trapping, with an application. *Biodiversity and Conservation*, in press.

Daily, G., Ehrlich, P., and Haddad, N. (1993). Double keystone bird in a keystone species complex. *Proc. Nat. Acad. Sci. USA*, **90**, 592–4.

Diamond, J. (1987). Extant unless proven extinct? Or extinct unless proven extant? *Conserv. Biol.*, **1**, 77–9.

Diamond, J. (1991). *The rise and fall of the third chimpanzee*. Radius, London.

Drake, J. *et al.* (1989). *Biological invasions*. Wiley, Chichester.

Ehrlich, P. R. (1964). Some axioms of taxonomy. *Syst. Zool.*, **13**, 109–23.

Ehrlich, P. R. (1984). The structure and dynamics of butterfly populations. In *The biology of butterflies*, (ed. R. I. Vane-Wright and P. R. Ackery), pp. 25–40. Academic Press, London.

Ehrlich, P., and Daily, G. (1993). Population extinction and saving biodiversity. *Ambio*, **22**(2–3), 64–8.

Ehrlich, P. R., and Ehrlich, A. H. (1981). *Extinction*. Random House, New York.

Ehrlich, P. R., and Ehrlich, A. H. (1990). *The population explosion*. Simon and Schuster, New York.

Ehrlich, P. R., and Ehrlich, A. H. (1991). *Healing the planet*. Addison-Wesley, New York.

Ehrlich, P. R., and Ehrlich, A. H. (1992). The value of biodiversity. *Ambio*, **21**, 219–26.

Ehrlich, P. R., and Holdren, J. P. (1971). Impact of population growth. *Science*, **171**, 1212–7.

Ehrlich, P. R., and Murphy, D. D. (1987). Conservation lessons from long-term studies of checkerspot butterflies. *Conserv. Biol.*, **1**, 122–31.

Ehrlich, P. R., and Murphy, D. D. (1988). Plant chemistry and host range in insect herbivores. *Ecology*, **69**, 908–9.

Ehrlich, P. R., Ehrlich, A. H., and Holdren, J. P. (1977). *Ecoscience: population, resources, environment*. W. H. Freeman, San Francisco.

Ehrlich, P. R., Murphy, D. D., Singer, M. C., Sherwood, C. B., White, R. R., and I. L. Brown (1980). Extinction, reduction, stability and increase: the responses of checkerspot butterfly (*Euphydryas*) populations to the California drought. *Oecologia*, **46**, 101–5.

Futuyma, D. J. (1991). Evolution of host specificity in herbivorous insects: genetic, ecological, and phylogenetic aspects. In *Plant-animal interactions: evolutionary ecology in tropical and temperate regions*, (ed. P. W. Price, T. M. Lewinsohn, G. W. Fernandes, and W. W. Benson), pp. 431–54. Wiley, New York.

Heath, J., Pollard, E., and Thomas, J. (1984). *Atlas of butterflies in Britain and Ireland*. Viking, New York.

Holdren, J. P. (1991). Population and the energy problem. *Pop. Environ.*, **12**, 231–55.

Holdren, J. P., and Ehrlich, P. R. (1974). Human population and the global environment. *Amer. Sci.*, **62**, 282–92.

Imhoff, M. L. (1993). The dependence of synthetic aperture radar backscatter on forest structure and biomass: potential application for global carbon models. Ph.D. thesis, Stanford University.

Kudrna, O. (1986). *Aspects of the conservation of butterflies in Europe*, Vol. 8. *Butterflies of Europe*, (ed. O. Kudrna). Aula, Weisbaden.

Kummer, D. M. (1991). *Deforestation in the postwar Philippines*. University of Chicago Press.

Lipscomb, C. G., and Jackson, R. A. (1964). Some considerations on some present day conditions as they affect the continued existence of certain butterflies. *Entomol. Rec. J. Var.*, **76**, 63–8.

Lowe-McConnell, R. H. (1987). *Ecological studies in tropical fish communities*. Cambridge University Press.

Martin, P. S., and Wright, H. E. (ed.) (1967). *Pleistocene extinctions*. Yale University Press, New Haven.

May, R. M. (1988). How many species are there on Earth? *Science*, **241**, 1441–9.

Myers, N. (1989). *Deforestation rates in tropical forests and their climatic implications*. Friends of the Earth, London.

Paine, R. (1966). Food web complexity and species diversity. *Amer. Nat.*, **100**, 65–75.

Rabinowitz, D., Cairns, S., and Dillon, T. (1986). Seven forms of rarity and their frequency in the flora of the British Isles. In *Conservation biology*, (ed. M. E. Soulé), pp. 182–204. Sinauer, Sunderland, MA.

Raup, D. M. (1988). Diversity crises in the geological past. In *Biodiversity*, (ed. E. O. Wilson), pp. 51–7. National Academy Press, Washington, DC.

Scott, J. A. (1986). *The butterflies of North America*. Stanford University Press.

Sibitani, A. (1989). Decline and conservation of butterflies in Japan. In *Decline and conservation of butterflies in Japan*, Vol. I, (ed. E. Hama, M. Ishii, and A. Sibitani), pp. 16–22. Lepidopterological Society of Japan, Osaka.

Simon, J., and Wildavsky, A. (1993). Facts, not species, are periled. *New York Times*, 13 May.

Smith, F., May, R., Pellew, R., Johnson, T., and Walter, K. (1993a). Estimating extinction rates. *Nature*, **364**, 494–6.

Smith, F., May, R., Pellew, R., Johnson, T., and Walter, K. (1993b). How much do we know about the current extinction rate? *Trends Ecol. Evol.*, **8**, 375–8.

Svensson, B. (1985). Local extinction and re-immigration of whirligig beetles (Coleoptera, Gyrinidae). *Ecology*, **66**, 1837–8.

Swales, S., and West, D. W. (1991). Distribution, abundance and conservation status of native fish in some Waikato streams in the North Island of New Zealand. *J. Roy. Soc. New Zealand*, **21**, 281–96.

Terborgh, J. (1983). *Five New World primates: a study in comparative ecology*. Princeton University Press.

Thomas, J. and Webb, N. (1984). *Butterflies of Dorset*. Dorset Natural History and Archeological Society, Dorchester.

UN (United Nations) (1988). *Statistic yearbook*. United Nations, New York.

Vitousek, P., Ehrlich, P., Ehrlich, A., and Matson, P. (1986). Human appropriation of the products of photosynthesis. *BioScience*, **36**, 368–73.

WCED (World Commission on Environment and Development) (1987). *Our common future*. Oxford University Press.

Wilson, E. O., (1992). *The diversity of life*. Harvard University Press, Cambridge, MA.

WRI (World Resources Institute) (1989). *World resources 1989–90*. Oxford University Press.

WRI (World Resources Institute) (1990). *World resources 1990–91*. Oxford University Press.

WRI (World Resources Institute) (1992). *World resources 1992–93*. Oxford University Press.

Index